TRANSMITTER BIOCHEMISTRY
OF
HUMAN BRAIN TISSUE

Proceedings of the Symposium held at the
12th CINP Congress, Göteborg, Sweden
June, 1980

Edited by

PETER RIEDERER

*Institute of Clinical Neurobiology,
Vienna*

and

EARL USDIN

*National Institute of Mental Health,
Rockville, Maryland*

This book was edited by Dr Usdin in
his private capacity. No official support
or endorsement by the NIMH is intended or
should be inferred.

First published 1981 by
Scientific and Medical Division
MACMILLAN PUBLISHERS LTD
London and Basingstoke
Companies and representatives throughout the world

ISBN 978-1-349-05934-8   ISBN 978-1-349-05932-4 (eBook)
DOI 10.1007/978-1-349-05932-4

Contents

Participants

Dr Anne C. Andorn
 Department of Pharmacology
 Case Western Reserve University
 School of Medicine
 Cleveland, Ohio 44106

Dr Edward Bird
 Department of Neuropathology
 McLean Hospital
 Belmont, Maryland 02178

Dr Walter Birkmayer
 Ludwig Boltzmann Institut
 fur Neurochemie
 Neurologische Abteilung
 Versorgungsheimplatz 1
 A-1130 Vienna Lainz, Austria

Dr David M. Bowen
 Department of Neurochemistry
 Institute of Neurology
 Queen Square
 London WC1N 3BG, UK

Dr Tim J. Crow
 Division of Psychiatry
 MRC Clinical Research Centre
 Watford Road
 Harrow, Middlesex HA1 3UJ, UK

Dr Piers Emson
 MRC Neurochemical Pharmacology
 Unit, Department of Pharmacology
 Medical School, Hills Road
 Cambridge CB2 2QD, UK

Dr E. S. Garnett
 McMaster Division
 Shedoke McMaster Hospital
 Box 2000, Station A
 Hamilton 28N 325, Ontario, Canada

Dr Carl-Gerhard Gottfries
 Psychiatric Research Center
 St. Joergen's Hospital
 S-422 03 Hisings Backa, Sweden

Dr Oleh Hornykiewicz
 Institute of Biochemical
  Pharmacology
 University of Vienna
 Borschkegasse 8a
 A-1090 Vienna, Austria

Dr Kenneth G. Lloyd
 Synthelabo-L.E.R.S.
 31, Avenue Paul Valliant Couturier
 92220 Bagneux, France

Professor Toshiharu Nagatsu
 Tokyo Institute of Technology
 4259 Nagatsuda, Midori-ku
 Yokohama, 227, Japan

Dr Peter Riederer
 Ludwig Boltzmann Institute for
 Clinical Neurobiology
 Lainz Hospital Pav. XI
 Wolkersbergenstrasse 1
 A-1130 Vienna, Austria

Dr U. D. Rinne
 Department of Neurology
 University of Turku
 SF-20520 Turku 52, Finland

Professor Merton Sandler
 Department of Chemical Pathology
 Bernhard Baron Memorial Research
  Laboratories
 Queen Charlotte's Maternity Hospital
 Goldhawk Road
 London W6 0XG, UK

Dr Philip Seeman
  Department of Pharmacology
  University of Toronto
  Toronto 5, Ontario, Canada

Dr Earl Usdin
  Neurosciences Research
  Branch, National Institute of
  Mental Health
  560 Fishers Lane — Room 9C-09
  Rockville, Maryland 20857

Preface

Research using human post-mortem brain tissue is performed in a number of scientific centers in the world; many find such research not only necessary but also invaluable to answer questions not only of basic neurochemistry but also of applied biochemistry, biochemical pharmacology and drug treatment.

The possibilities and limitations of neurochemical research using human brain tissue have not been discussed previously *in extenso*; a number of questions are still unresolved. Although there has recently been a 'Workshop on Dissection of Human Brain Tissue' (Vienna, 1979), a conference where research groups working in the fields of neurochemistry and pharmacology of the human brain exchanged results and techniques had not been organized prior to the 1980 CINP Congress in Göteborg, Sweden. At that meeting, there was a Symposium on Transmitter Biochemistry of Human Brain Tissue, with discussion extended to the molecular level for diseases of the central nervous system.

Methods are discussed in this volume for biochemical analyses using post-mortem brain tissue, including how subcellular fractions (mitochondria, synaptosomes, membranes, etc.) behave in unfrozen tissue in comparison to frozen tissue and how post-mortem time, age, disease, therapy, preparation of homogenates, storage conditions, etc. influence results. Such methodological questions need to be discussed in the light of comparison between neuropathology and neurochemistry. Studies on uptake, release and receptor binding are also covered. Comparisons are made of data obtained with human brain tissue with data from animal experiments.

Animal models have their limitations in simulating psychiatric and neurological disorders; studies on schizophrenia, Parkinson's Disease, Huntington's Disease, senile dementia of Alzheimer's type, and problems of the aging brain contribute to a better understanding of the underlying pathomechanisms of such disorders and may help in developing new therapeutic strategies.

This interchange of ideas and experience by international experts allows us to take stock of knowledge on the neurochemistry of human brain tissue. We hope this important information will stimulate basic research as well as applied pharmacology.

It is now 20 years since, as a consequence of analyses of human brain tissue, a therapy was found for a neurological disease: L-Dopa therapy for Parkinson's Disease. We are happy that Dr Birkmayer and Dr Hornykiewicz, both responsible for opening this door to human brain biochemistry and pharmacology, have accepted our invitation to contribute to this Symposium.

Peter Riederer, Earl Usdin

Abbreviations

ACE, angiotensin converting enzyme
ADTN, 2-amino-6, 7-dihydroxy-1, 2,3,4-tetrahydronaphthalene
AMP, adenosine monophosphate
ATD, Alzheimer's-type dementia
ATPase, adenosine triphosphatase
BBB, blood—brain barrier
BGP, brain gastric peptide
CA, catecholamine(s)
cAMP, cyclic AMP
CAT, choline acetyltransferase
CBF, cerebral blood flow
CCK, cholecystokinin
CNS, central nervous system
COMT, catechol O-methyltransferase
CSF or c.s.f., cerebrospinal fluid
DA, dopamine
DBH, dopamine β-hydroxylase
DHEC, dihydroergocryptine
DNA, deoxyribonucleic acid
Dopa or dopa or DOPA, dihydroxy-phenylalanine
DOPAC, dihydroxyphenylacetic acid
EDTA, ethylenediamine tetraacetic acid
GABA, γ-aminobutyric acid
GABA-T, GABA transaminase
GAD, glutamic acid decarboxylase
GC, gas chromatography
GRH, gonadotropin-releasing hormone
H.D., Huntington's Disease

5-HIAA, 5-hydroxyindole acetic acid
HPLC, high performance (or pressure) liquid chromatography
^{3}HSP, ^{3}H-spiroperidol
5-HT or HT, serotonin
5-HTP, 5-hydroxytryptophan
HVA, homovanillic acid
KYN, kynurenine
MAO, monoamine oxidase
MC, magnocellular
met, methionine
MHPG, 3-methoxy-4-hydroxyphenyl-alanine
NA, noradrenaline (or norepi-nephrine)
NM, normetanephrine
NTR, neurotransmitter receptor
pat., patients
PC, parvocellular
rCBF, regional CBF
RN, red nucleus
RNA, ribonucleic acid
SN, substantia nigra
TH, tyrosine hydroxylase
TRP, tryptophan
TRP-OH, TRP hydroxylase
TYR, tyrosine
TYR-OH, TYR hydroxylase
VAL, valine
VIP, vasoactive intestinal polypeptide

Section I
Behavior and Biochemistry:
An Introductory Essay

Behavior and Biochemistry: An Introductory Essay

KONRAD LORENZ

Österreichische Akademie der Wissenschaften,
Institut für Vergleichende Verhaltensforschung,
A-3422 Altenberg, Austria

It is questionable whether someone with next-to-no understanding of the content of this book should be permitted to write its preface. I can only justify this introduction by saying that one can well estimate the importance of a new research field and its results if one is able to assess the gap in our present knowledge which they help to bridge.

Many neurophysiologists and behaviorists underestimate the width of the abyss which lies between our knowledge of behavior and its underlying neurophysiology. Even if one believes in the outmoded idea that the "reflex" is the basic element of all animal and human behavior, and even if one thinks that "the nervous system of animals is nothing else but an organ for the processing and transport of information" (Neureiter, 1974) a tremendious cleft still remains between the action of a single neural element and the behavior of the whole organism. We have learnt from Erich von Holst, Kenneth Roeder and others that the neuron itself, indeed even an axon, has a complex and comprehensive inventory of spontaneous behavior. We know, too, that the reaction of an effector cell to a stimulus is described not by a single impulse but by a modulation of the frequency of continuously and spontaneously emitted impulses. As there is so much complexity at even this "simple" level, the gap between the behavior of an organism and that of its constituent elements appears much greater. Thus the research strategy of our best neuroethologists (such as F. Huber, and W. Heiligenberg) is to study relatively simple behavior in relatively simple organisms. The gap between our present understanding of elementary neuronal and sensory functions and our ability to interpret animal and human behavior is tremendous, but it must be emphasized that it can, in principle, be bridged by the knowledge of the functions of elements and of the s t r u c t u r e s   into which they are combined.

There is, however, another "hiatus" - as Nicolai Hart-mann has described it. It is the inability, in our present way of thinking to overcome the basic problem of the body-mind relationship. This is in principle insoluble since the relationship between the physio-logical and the subjective event is, as expressed by Max Hartmann, alogical. We know that some physiologi-cal processes are generally accompanied by certain subjective phenomena, while other such processes are not consciously experienced. This relation of the physiological with the subjective is not correlated with the complexity of the process. Some simple events, such as the inflammation of tooth-pulp, have an intense subjective parallel, while some highly complex pheno-mena, for example computations performed by our Gestalt perception, are not only unconscious but not even accessible to introspection. Thus the old idea that simple neurosensory processes are "merely" physiologi-cal, while higher, more complex events are better explained psychologically (found in the otherwise excellent animal psychology textbook by Hempelmann) is totally false.

Provided one assumes the inherent insolubility of the body-mind problem, there are three possible attitudes that can be taken with regard to it. Firstly, one may assume a reciprocal interaction between the physiolo-gical processes and the concomittant subjected experi-ences. This I refuse to accept because it postulates a non-physical causation of physical events. The second assumption is that a prestabilized harmony or a parallel exists between psychological and physio-logical events. There are, however, limitations to "psycho-physiological parallels", since there are so very many neuro-physiological functions which do not have any subjective parallels. The third assumption is that body and mind are only two different aspects of the same reality, each defined by a different cognitive apparatus of our brain. This "identity doctrine" meets the same difficulties as that of parallelism. There is no doubt that only a few particular neurological events are accompanied by

4

objective phenomena. They rise above the threshold of our consciousness as does the tip of an iceberg, while its greater part remains below the surface.

In spite of this difficulty I adhere to it for the simple reason that the cognitive structures of every normal human being are dictating it. When I assert that my fried B. is sitting beside me, I mean neither the presence of his physical system accessible to physiological research, nor do I mean the unity of all his undisputed mind and his subjective phenomena, so similar to my own. What I mean is quite exactly the unity of both.

One can step aside from this problem by neglecting subjective reaction and by examining only objectively measurable animal behavior. This is the approach of the large American school of behaviorism. This narrowing of interest is justifiable when studying either simple organisms or specific and relatively simple learning mechanisms. However, in the case of man and the higher animals this exclusion of subjective experience is the greatest sin in our search for truth: as it means the refusal to accept perfectly accessible knowledge.

If one is convinced of the fundamental unity of the organisms, then it is justifiable to correlate qualitative experience with neurophysiological action. Of course, the reverse is not necessarily true. But every experience, even the most transient dream, correlates with a physiological event. There are no miracles! Even when this assertion is more religious than scientific it is nevertheless the basis of the scientific study of man.

Thus it is perhaps justifiable to say that the nervous system is more than just an organ for the processing and transport of information. Indeed, since "I" am my own nervous system, I contradict with greatest possible assurance the statement that I am "nothing else but an information processing and transport system".

Both complex and basic processes of our brain are accompanied by subjective experience. Indeed, regulatory processes occuring in the brain are able to influence our psyche to a great extent. This is important in relation to this publication, as without the observation of subjective phenomena (so despised by behaviorists) it would hardly have been possible to have advanced so far in our understanding of the biochemical function of the brain stem. It raises our confidence in that such observation of subjective phenomena may be a reliable indicator of the excess as well as of the deficiency of certain biochemical trans- mitter substances, in particular in brain stem regions. The importance of this has been illustrated by the fact that through such observations and subsequent research it has been possible to correct abnormally functioning regulatory mechanisms by chemotherapy.

If it is at all possible to bridge the gap between biochemical function and behavior, then the work on transmitter abnormalities in the brain stem will be seen to be a vital contribution.

Section II
Basic Techniques

Importance of Topographic Neurochemistry in Studying Neurotransmitter Systems in Human Brain: Critique and New Data

OLEH HORNYKIEWICZ

Institute of Biochemical Pharmacology,
University of Vienna, Austria

INTRODUCTION

Neurotransmitter studies in human brain obtained at autopsy are limited by many factors. Among these are: age of the patient; disease state (acute, chronic); drug history; immediate pre-mortem status (sudden death, prolonged coma; metabolic decompensation); and autopsy time (i.e. time elapsed between death and freezing of the brain). In addition to these "external" factors, the regionally uneven distribution of many putative neurotransmitter substances represents an "internal" uncertainty factor that can grossly distort the results. Therefore, meticulous care has to be taken in defining and dissecting out as accurately as possible the individual brain nuclei under study. In the following discussion the importance of accurate topographical localization will be illustrated by data taken from the published literature on regional distribution of dopamine (DA), noradrenaline (NA) and serotonin (5-HT) in the human brainstem and basal ganglia. In addition, our data obtained in a standardized topographical study of DA and NA distribution within the caudate-putamen-accumbens complex will be discussed.

MONOAMINE LEVELS IN SELECTED REGIONS OF THE HUMAN BRAIN
RESULTS FROM DIFFERENT LABORATORIES

The purpose of the following selected literature survey is to show that even for anatomically well-defined brain regions the data on monoamine levels differ considerably from laboratory to laboratory; in areas of doubtful anatomical delineation these differences often cast justified doubts regarding the anatomical identity of the analyzed region. To illustrate these points, the following brain regions have been selected for discussion: (a) areas that form well-defined nuclei, both gross-anatomically and morphologically, such as caudate-putamen and red nucleus; (b) gross-anatomically ill-delineated regions composed of several morphologically distinct sub-nuclei, such as the raphe region and the hypothalamus; and (c) gross-anatomically and morphologically not yet defined nuclei, such as the nucleus accumbens as a separate entity within the striatal nuclear complex.

Anatomically Well-Defined Nuclei

   Caudate nucleus and putamen. Table 1 shows the DA concentrations in the caudate nucleus and putamen of control subjects as reported by different laboratories. It is quite obvious that the striatal DA levels measured in the various laboratories vary widely. For the caudate nucleus the range is more than 10-fold: from 0.42 ug/g reported by Mackay et al.(1978) to 4.94 ug/g found by Moses and Robins (1975); for the putamen there is a similar range: from 1.32 ug/g in the study of Winblad et al. (1979) to 5.16 ug/g in the study of Moses and Robins (1975) (Mackay et al.did not report on DA in putamen). Not all authors reported on autopsy time and age of patients; from the available data (cf. Table 1) it seems obvious that in the above studies both factors played a role in determining the levels of DA. This is in accord with the conclusion reached by Carlsson and Winblad in an earlier study (1976). For the studies shown in Table 1, autopsy time seemed to be of greater importance than age of the patients. Thus, in cases with a mean autopsy time less than 24 h the DA concentration in the caudate nucleus was around the 4 ug/g mark (Carlsson and Winblad,1976; Lloyd et al., 1975; Moses and Robins, 1975); in the other studies with autopsy times longer than 24 h the caudate DA levels were invariably less than 2 ug/g (Adolfsson et al., 1979a; Bird et al., 1977; Crow et al., 1979; Mackay et al., 1978; Winblad et al., 1979) This strong influence of autopsy time casts some doubt on the validity of conclusions drawn from biochemical studies performed in laboratory animals under conditions designed to imitate the human brain postmortem conditions (e.g. Spokes and Koch, 1978). That the autopsy time is not the full explanation for the

10

TABLE 1. Dopamine in Human Brain (ug/g) - Caudate Nucleus and Putamen*

	Age (yrs)	Autopsy time (h)	Caudate nucleus	Putamen
Mackay et al., 1978	61	30	0.42 ± 0.07 (9)	---
Winblad et al., 1979	79	33	1.33 ± 0.53** (9)	1.32 ± 0.61** (9)
Rinne and Sonninen, 1973	?	12-48	1.37 ± 0.23 (13)	2.34 ± 0.45 (12)
Adolfsson et al., 1979a	61	27	1.57 ± 0.97** (23)	1.45 ± 1.10** (24)
Crow et al., 1979	76	48	1.6 ± 0.3 (19)	2.0 ± 0.4 (19)
Bird et al., 1977 (1979)	60	39	(1.73 ± 0.13) (51)	2.20 ± 0.23 (29)
Carlsson and Winblad, 1976	60	14	3.83 (6)	---
Birkmayer et al., 1974	?	?	3.85 ± 0.60 (9)	4.18 ± 0.74 (9)
Lloyd et al., 1975	68	13	4.06 ± 0.47 (18)	5.06 ± 0.39 (17)
Moses and Robins, 1975	48	10	4.94 ± 0.83 (7)	5.16 ± 0.75 (7)

*wet weight ± s.e.m. (number of estimated cases in parentheses); ** s.d.

11

wide discrepancies in respect to DA levels in the
human caudate and putamen is suggested by the very low
caudate DA concentration found by Mackay et al. (1978)
in a group of patients with a mean autopsy time of
30 h; in this respect, several other studies shown in
Table 1 measured 3 - 4 fold higher caudate DA levels
in patients with autopsy times ranging from 27 to 48 h.
In their survey of literature Adolfsson et al. (1979a)
suggested that the variations of DA levels in the
human material may be due to differences in methodology
of DA determination as performed in different laboratories. However, a cursory survey of studies on caudate DA in the rat reported from different laboratories
employing a variety of different extraction and detection methods, reveals a markedly narrower range (7-
13 ug/g)(e.g.: Brownstein et al., 1974; Koslow et al.
1974: Roizen et al., 1976; St. Laurent et al., 1975; Tassin et
al., 1976; Versteeg et al., 1976). Another explanation for the
wide variation could be the possibility of an uneven pattern of
the dopaminergic innervation of the human striatum, e.g. analogous
to that reported by Tassin et al. (1976) for the rat caudate
nucleus. As shown in Table 2, in the human caudate and putamen
such differencies are quite small compared with the disproportion-
ately wide variation of the DA levels shown in Table 1.

TABLE 2. Dopamine in Human Brain - Caudate and Putamen[*]

Caudate nucleus
head - inferior	3.33 ± 0.72	(6)		
- superior	3.43 ± 0.85	(6)		
tail - rostral	3.12 ± 0.91	(6)	Fahn et al.	
- caudal	0.65 ± 0.22	(6)	1971	
anterior	1.61 ± 1.02	(18)[**]		
posterior	1.54 ± 1.10	(18)[**]		
Putamen			Adolfsson	
anterior	1.43 ± 1.23	(18)[**]	et al.,1979	
posterior	1.91 ± 1.40	(18)[**]		

[*] values are in ug/g wet weight ± s.e.m. (number of
estimated cases in parentheses); [**] s.d.

DA levels lower than 1 ug/g were found by Fahn et al.
(1971) only in the caudal tail of the human caudate
nucleus, with all other parts containing DA concen-

trations higher than 3 ug/g (see Table 2); it is
highly improbable that any of the studies shown in
Table 1 included such a minute and not easily accessible portion of the caudate as the caudal tail. However, considering the slighty uneven DA pattern within the striatal nuclei it may be of crucial importance
when attempts are made to interpret the significance
of only slightly abnormal values in pathological
material, such as the somewhat higher DA levels measured in the caudate nucleus and putamen (Crow et al.,
1979) and nucleus accumbens (Bird et al., 1977) of
schizophrenic patients.

Red nucleus. Table 3 shows the results of 6 studies
from different laboratories in which the DA levels in
the red nucleus were determined. It is obvious that
the results can be divided into 2 groups: those 3
studies which found DA levels around the 0.2 ug/g mark
(Bertler, 1961; Moses and Robins, 1975) or below the
level of sensitivity of the assay procedure (Farley et
al., 1980); and studies reporting on DA concentrations
around the 1 ug/g mark (Jellinger and Riederer,1977;
Birkmayer et al., 1974 - both studies are from the
same laboratory; Sano et al., 1959).

TABLE 3. Dopamine in Human Brain - Red Nucleus and
Substantia Nigra

	Red nucleus*	Subst.nigra*
Bertler, 1961	0.19 (4)	0.40 (4)
Moses and Robins, 1975	0.21 ± 0.08 (5)	0.95 ± 0.28 (5)
Farley et al., 1980	0.25 (8)	0.95 ± 0.13 (11)
Jellinger and Riederer, 1977	0.76 ± 0.041 (5)	0.63 ± 0.04 (5)
Birkmayer et al., 1974	0.89 ± 0.14 (9)	0.71 ± 0.14** (9)
Sano et al., 1959	1.17 (3)	0.38 (3)

* values are in ug/g wet weight ± s.e.m. (number of
estimated cases in parentheses); ** caudal part

The high red nucleus DA levels of the latter authors
are surprising because, as shown in Table 3, in the
same studies the substantia nigra had DA levels which
were in the same range, or lower than in the red
nucleus. The substantia nigra has been established as
a major dopaminergic brain region in all species
examined, being the source of the dense dopaminergic
innervation of the striatum. In contrast, no specific
catecholamine fluorescence has so far been detected in
the red nucleus of laboratory animals (rat: Dahlström and Fuxe, 1964; Fuxe, 1965; Lindvall and Björklund, 1978; Palkovits and Jacobowitz, 1974 - cat:
Poitras and Parent, 1978 - dog: Shimada et al., 1976 -
primate (pigmy marmoset):Jacobowitz and MacLean, 1978).
Since the red nucleus in the human brain is a well-
defined anatomical region, and gross contamination
with substantia nigra tissue can be practically ruled
out, there must be other (possibly including methodological) reasons for these large discrepancies. A study
analyzing separately the two major components of the
red nucleus (the magnocellular and parvocellular
parts) may possibly help to resolve this question.

Gross-Anatomically Ill-Defined Regions Composed of
Morphologically Distinct Sub-Nuclei
    The raphe region of the lower brain stem. Table 4
summarizes the results of 2 studies on the 5-HT levels
in the raphe region of the lower brain stem in control
subjects. Riederer and Wuketich (1976) in their study
"...cut out, starting from the rhombencephalon, from
the pons and medulla oblongata a stip of tissue aver-
aging 3 mm which, however, contained white and grey
substance of the reticular formation mixed indiscrimi-
nately". In this strip of "raphe" tissue the authors
measured between 0.51 and 0.56 ug/g 5-HT. Lloyd et al.
(1974) on the other hand made the attempt at isolating
the individual raphe nuclei with the help of the
atlas of Olszewski and Baxter (1954) and the study
published by Braak (1970); since in the human brain
there is gross-anatomically no clearly visible delineation of the raphe nuclei, it has to be assumed
that in this study there was some contamination of the
individual nuclei with adjoining tissue. It can be
seen in Table 4 that in the study of Lloyd et al.(1974)
except for the nucleus raphe pallidus all other raphe
nuclei contained 5-HT levels (between 1.1 and 2.3 ug/g)
considerably higher that the value reported by Riederer and Wuketich (1976). It is highly improbable that
the latter authors' "raphe" tissue was composed of the

nucleus raphe pallidus; this nucleus forms a small, inconspicuous aggregation of nerve cells in not easily accessible deeply located parts of the lower medulla.

TABLE 4. Serotonin in Human Brain - The Raphe Region[*]

"Raphe"	0.51 - 0.56	(27)	Riederer and Wuketich,1976
N.raphe dorsalis	2.22 ± 0.13	(5)	
N.raphe centralis sup.	2.25 ± 0.19	(5)	
N.raphe pontis	1.34 ± 0.21	(5)	Lloyd et al.,
N.raphe centralis inf.	1.32 ± 0.12	(5)	1974
N.raphe obscurus	1.07 ± 0.14	(5)	
N.raphe pallidus	0.61 ± 0.09	(5)	

[*] values are in ug/g wet weight ± s.e.m. (number of estimated cases in parentheses)

Thus it has to be concluded that in this study, the tissue defined as "raphe" did not include major portions of any of the well-known and easily accessible raphe nuclei.

Hypothalamus. Table 5 shows the results of 5 studies selected from the literature reporting on NA levels in the hypothalamus of control subjects.It can be seen that, as with the other brain regions, there also was a wide variation of hypothalamic NA concentrations as determined in different laboratories, ranging between 0.51 ug/g in the study of Mackay et al. (1978) and 1.83 ug/g reported by Farley et al. (1978a).

TABLE 5. Noradrenaline in Human Brain - Hypothalamus[*]

Mackay et al., 1978	0.51 ± 0.15	(7)
Adolfsson et al., 1979b	0.75 ± 0.40[**]	(10-20)
Winblad et al., 1979	0.86 ± 0.39[**]	(10)
Pare et al., 1969	0.98 ± 0.07	(15)
Farley et al., 1978a	1.83 ± 0.18	(12)

[*] values are in ug/g wet weight ± s.e.m. (number of estimated cases in parentheses); [**] s.d.

These differences could be partly due to different
autopsy intervals, with the study of Farley et al.
(1978a) including material with the shortest mean
autopsy time (16.5 h). Although being an easily accessible brain region, the hypothalamus, on the other
hand, is anatomically heterogeneous, consisting of a
series of individual sub-nuclei. It is thus possible
that the differences in NA levels reported from
different laboratories reflect this heterogeneity of
the area with different sub-nuclei containing different
amounts of NA. In this respect, Table 6 summarizes the
results of 2 studies on NA in which an attempt   was
made at subdividing the hypothalamus into its components. In the study of Farley and Hornykiewicz(1977)
the hypothalamus was divided into the anterior, posterior and lateral portion; Moses and Robins (1975), on
the other hand, performed a detailed dissection of
individual nuclei. As can be seen from Table 5, the
results of the two studies agree fairly well with
each other. Farley et al. (1977) measured the highest

TABLE 6. Noradrenaline in Human Brain - Subdivisions
of Hypothalamus*

Hypothalamus				
anterior	1.83 ± 0.25	(4)	Farley and	
posterior	1.31 ± 0.29	(4)	Hornykiewicz,	
lateral	1.25 ± 0.09	(8)	1977	
N.ant.med.dor.	1.87 ± 0.38	(4)		
N.ant.med.vent.	1.26 ± 0.18	(6)		
N.ant.lat.dor.	1.74 ± 0.47	(5)		
N.ant.lat.vent.	1.12 ± 0.22	(5)	Moses and	
N.post.med.dor.	1.44 ± 0.08	(5)	Robins,	
N.post.med.vent.	0.94 ± 0.27	(8)	1975	
N.post.lat.dor.	1.06 ± 0.33	(4)		
N.post.lat.vent.	0.95 ± 0.46	(3)		

*values are in ug/g wet weight ± s.e.m. (number of
estimated cases in parentheses).

NA concentrations(1.83 ug/g) in the anterior part of
the hypothalamus and the lowest (1.25 ug/g) in the
lateral subdivision, with the posterior hypothalamus
having intermediate NA concentrations. In the study
of Moses and Robins (1975), the NA levels in the individual nuclei ranged from 0.94 ug/g in the posterior

16

medial ventral nucleus to 1.87 ug/g in the anterior
medial dorsal nucleus. Thus, the studies reporting on
hypothalamic NA levels below the 1 ug/g mark (cf.
Table 5) may have included a varying number of nuclei
containing less NA than the NA rich anterior hypothalamic subdivisions.

Gross-Anatomically and Morphologically not yet Defined Nuclei

Nucleus accumbens (septi). Table 7 summarizes the
data selected from the literature on DA and NA in the
human nucleus accumbens. In these studies, the DA
concentration in the nucleus accumbens showed a wide
variation, ranging from 0.77 ug/g (Mackay et al.,
1978) to 3.68 ug/g (Farley et al. 1978b).

TABLE 7. Dopamine and Noradrenaline in Human Brain –
Nucleus Accumbens

	Dopamine*	Noradrenaline*
Mackay et al., 1978	0.77 ± 0.29 (5)	0.30 ± 0.24 (4)
Crow et al., 1979	0.9 ± 0.3 (18)	0.13 ± 0.03 (10)
Bird et al., 1979	1.22 ± 0.10 (46)	0.13 ± 0.01 (40)
Jellinger and Riederer, 1977	1.80 ± 0.12 (9?)	---
Farley et al., 1978b	3.68 ± 0.73 (9)	1.58 ± 0.16 (8)

* values are in ug/g wet weight ± s.e.m. (number of
estimated cases in parentheses)

On the whole, this variation was of the same magnitude
as the variation reported by the same laboratories
for DA in the caudate nucleus and putamen (cf. Table 1;
studies of Farley et al., 1978b, Table 7, and Lloyd et
al., 1974, Table 1, are from the same laboratory).
Thus, it could be assumed that the above differences
for the nucleus accumbens are due to the same factors
as those discussed for the caudate and putamen (e.g.
autopsy time, etc.; see above). However, this explanation is made rather unlikely if one considers the

NA levels in what the above studies define as nucleus accumbens. As shown in Table 7, the first 3 authors measured very low NA levels (0.13 - 0.3 ug/g) in this subdivision of the striatum. In sharp contrast, Farley et al. (1977; 1978a, 1978b) found quite high NA levels (up to 1.58 ug/g) in their nucleus accumbens. This very large discrepancy clearly points to basic differences between the different laboratories regarding the ana-tomical definition of the area involved.

DETAILED TOPOGRAPHIC STUDY OF DOPAMINE AND NORADRENA-LINE IN THE HUMAN CAUDATE-PUTAMEN-ACCUMBENS COMPLEX: PRELIMINARY RESULTS

The wide and in many instances irreconcilable dis-crepancies between the results obtained in different laboratories on monoamine levels in human brain prompt-ed us to attempt a detailed topographic distribution study of the DA and NA in the caudate-putamen-accum-bens complex (Hörtnagl, Schlögl, Sperk, Hornykiewicz, in preparation). So far, brains of 3 female control subjects (age: 72, 38, 57 yrs; autopsy time: 6, 3.5, 8 hrs) have been analyzed. Six coronal sections (each appr. 3mm thick) were taken from frozen hemispheres, di-viding the part of the striatum situated rostrally to the anterior commissure (see Figure 1) (cf.Riley, 1960).

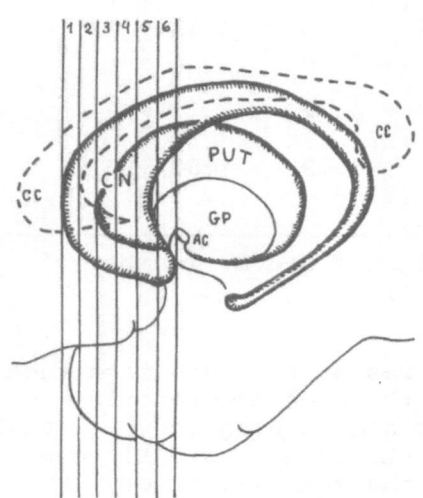

Figure 1. Diagram showing the level and plane of 6 sections through rostral striatum. Section 1: through rostral limit of caudate (CN); Section 2: through rostral limit of putamen (PUT); Section 3: through rostral part of junction of CN and PUT; Section 4: through caudal part of the junction area; Section 5: through rostral limit of globus pallidus (GP); Section 6: through anteri-or commissure (AC).

In all sections the caudate and putamen were subdivided into dorsal, intermediate and ventral portions; so was the junction area between caudate and putamen in sec-tions 3,4 and 5. (The tissue was homogenized in .1N per-

chloric acid, and partly purified by extraction with
ethylacetate and adsorption to aluminium oxide; DA and
NA were measured by means of the high performance liquid chromatography method with electrochemical detection).

Dopamine

Caudate nucleus. In the caudate nucleus, highest DA
concentrations were measured in the intermediate portions of sections 3 - 6 and the dorsal portion of part
5. The average concentration of DA in section 1, i.e.
the rostral pole of the nucleus, was somewhat lower
than in the other sections. Apart from this, there
was no apparent rostro-caudal DA gradient in the
analyzed (rostral) part of the caudate nucleus.

Putamen.In contrast to the caudate nucleus, the
putamen showed a clear rostro-caudal gradient, with DA
levels increasing in the caudal direction. This is in
accord with the earlier report of Adolfsson et al.
(1979a) (see also Table 2). At the level of sections
4 and 5 the ventral portions of the putamen tended to
have somewhat lower DA concentrations than the dorsal
and intermediate portions.

Junction area between caudate and putamen. At the
level of section 3 and 4 the DA concentration in the
junction area was very similar to that found in the
adjoining portions of the caudate and putamen. At the
level of section 5 (rostral limit of the globus
pallidus) the junction area of the striatal complex
contained distinctly lower DA levels than the surrounding parts of the caudate and putamen. In sections
3 and 5, the ventral subdivision of the junction area
was slightly richer in DA than the dorsal portions.

In conclusion, the above data on the detailed distribution pattern of DA within the human striatal
complex demonstrate the presence of DA gradients in
both the caudate nucleus and the putamen. However, the
DA distribution is,on the whole, not characteristic
enough to be helpful in answering the question as to
the exact topographical localization of the nucleus
accumbens in the human brain.

Noradrenaline

Caudate nucleus. All portions of the caudate nucleus
were poor in NA. In general, the NA levels were no
more than 1/100 of the DA levels. No gradients for NA
could be observed in this nucleus.

Putamen. In the putamen, the NA concentrations were,

as a rule, slightly higher than in the caudate nucleus, about 1/50 of the DA levels. There also was a distinct, though rather shallow, NA gradient discernable, with the caudal sections (especially their ventral subdivisions) containing NA levels up to the 0.2 ug/g (wet weight) mark.

Junction area between caudate and putamen. In sharp contrast to the very low NA levels in the caudate nucleus and putamen proper, the junction area of these nuclei, particularly in sections 4 and 5, had 10 - 40 fold higher NA levels, between 1.5 and 2 ug/g wet wieght; in this respect, the dorsal and intermediate portions of the junction area were considerably richer in NA than the ventral portion. Thus, the junction area between the caudate and putamen (the ventromedial striatum) can be counted among the regions containing the highest concentrations of NA measured in the human brain. These results confirm our earlier finding of high NA levels in this part of the human striatum (nucleus accumbens) using a different (transverse) plane of dissection (Farley et al., 1977; 1978a; 1978b).

Conclusion. The exact anatomical location of the nucleus accumbens in the human brain has not yet been established; however, all available sources agree that the nucleus accumbens is part of the ventral portion of the rostral striatum (cf. Brockhaus, 1942; Crosby et al., 1962). Our above study on the detailed distribution pattern of NA in the rostral striatum has produced evidence for the occurrence of particularly high NA concentrations in those portions of the striatum which constitute the junction area between the caudate and putamen. It is tempting to assume that the exceptionally high NA concentration in this striatal subdivision reflects an anatomically distinct feature of the striatal complex. In this respect, the idea suggests itself that this portion of the striatal complex may in fact be identical with the nucleus accumbens, with NA serving as a specific neurochemical marker for this region. The very low NA concentrations reported for the human "accumbens" by several authors (see Table 7) suggest that most likely they were dealing with a part of the ventral striatum anatomically different from what we should like to consider as the nucleus accumbens.

REFERENCES

Adolfsson, R., Gottfries, C.G., Roos, B.E. and Winblad B. (1979a).Post-mortem distribution of dopamine and homivanillic acid in human brain, variation related to age, and a review of the literature.
J.Neural Transmission 45, 81-105.

Adolfsson, R., Gottfries, C.G., Roos, B.E. and Winblad, B. (1979b).Changes in the brain catecholamines in patients with dementia of Alzheimer type.
Brit.J.Psychiatr. 135, 216-223.

Bertler, A. (1961). Occurrence and localization of
catecholamines in the human brain. Acta physiol.
scand. 51, 97-107.

Bird, E.D., Barnes, J., Iversen, L.L., Spokes, E.G.,
Mackay, A.V.P. and Sheperd, M. (1977). Increased
brain dopamine reduced glutamic acid decarboxylase
and choline acetyltransferase activity in schizophrenia and related psychoses. Lancet II, 1157-1159.

Bird, E.D., Spokes, E.G. and Iversen, L.L. (1979).
Brain norepinephrine and dopamine in schizophrenia
(technical comment). Science 204, 93-94.

Birkmayer, W., Danielczyk, W., Neumayer, E. and
Riederer, P. (1974). Nucleus ruber and L-dopa
psychosis: biochemical post-mortem findings. J.
Neural Transmission 35, 93-116.

Braak, H. (1970). Über die Kerngebiete des menschlich-
en Hirnstammes II. Die Raphekerne. Z.Zellforsch.
107, 123-141.

Brockhaus, H. (1942). Zur feineren Anatomie des Septum
und des Striatum. J.Physiol.u.Neurol.51, 1-56.

Brownstein, M., Saavedra, J.M., and Palkovits, M.
(1974). Norepinephrine and dopamine in the limbic
system of the rat. Brain Res. 79, 431-436.

Carlsson, A. and Winblad, B. (1976). Influence of age
and time interval between death and autopsy on
dopamine and 3-methoxytyramine levels in human
basal ganglia. J. Neural Transmission 38, 271-276.

Crosby, E.C., Humphrey, T. and Lauer, E.W. (1962).
Correlative Anatomy of the Nervous System. Macmillan
New York.

Crow. T.J., Baker, H.F., Cross,A.J., Joseph,M.H.,
Lofthouse,R., Longden, A., Owen, F., Riley, G.J.,
Glover, V., Killpack, W.S. (1979). Monoamine
mechanisms in chronic schizophrenia: post-mortem
neurochemical findings. Brit.J.Psychiatr.134,249-256

Dahlström, A. and Fuxe, K. (1964). Evidence for the
existence of monoamine-containing neurons in the
central nervous system. I.Demonstration of monoamines in the cell bodies of the brainstem neurons.

21

Acta physiol.scand. <u>62</u>, 1-55.
Fahn, S., Libsch, L.R. and Cutler, R.W.(1971). Mono-
    amines in the human neostriatum. Topographic di-
    stribution in normals and in Parkinson's disease
    and their role in akinesia, chorea and tremor.
    J.Neurol.Sci.<u>14</u>, 427-455.
Farley, I.J. and Hornykiewicz, O. (1977). Noradrenali-
    ne distribution in subcortical areas of the human
    brain. Brain Res. <u>126</u>, 53-62.
Farley, I.J., Price, K.S., McCullough, E., Deck, J.H.N.
    Hordynski, W. and Hornykiewicz, O. (1978a). Nor-
    epinephrine in chronic paranoid schizophrenia:
    above-normal levels in limbic forebrain. Science
    <u>200</u>, 456-458.
Farley, I.J., Price, K.S. and Hornykiewicz, O. (1978b).
    Monoaminergic systems in the human limbic brain. In
    <u>Limbic Mechanisms</u>, (eds. K.E.Livingston and O.
    Hornykiewicz), Plenum, New York. 333-349.
Farley, I.J., Shannak, K.S. and Hornykiewicz, O.(1980)
    Dopamine in brain stem nuclei. Manuscript in pre-
    paration.
Fuxe, K. (1965). Evidence for the existence of mono-
    amine neurons in the central nervous system.IV:
    Distribution of monoamine nerve terminals in the
    central nervous system. Acta physiol.scand.<u>64</u>,
    suppl.247, 39-85.
Jacobowitz, D.M. and MacLean, P.D.(1978). A brainstem
    atlas of catecholaminergic neurons and serotoner-
    gic perikarya in a pygmy primate (cebuella pygmaea).
    J.Comp.Neurol.<u>177</u>, 397-416.
Jellinger, K. and Riederer, P. (1977). Brain mono-
    amines in metabolic (endotoxic) coma. A preliminary
    biochemical study in human post-mortem material.
    J.Neural Transmission <u>41</u>, 275-286.
Koslow, S.H., Racagni, G. and Costa, E. (1974). Mass
    fragmentographic measurement of norepinephrine,
    dopamine, serotonin and acetylcholine in seven
    discrete nuclei of the rat tel-diencephalon. Neuro-
    pharmacol. <u>13</u>, 1123-1130.
Lindvall, O. and Björklund, A.(1978). Organization of
    catecholamine neurons in the rat central nervous
    system. In: <u>Handbook of Psychopharmacol. 9</u> (eds.
    L.L.Iversen, S.D.Iversen,S.H.Snyder).Plenum Publ.
    Corp.New York. 139-231.
Lloyd, K.G., Farley, I.J., Deck, J.H.N. and Horny-
    kiewicz, O. (1974). Serotonin and 5-hydroxyindole-
    acetic acid in discrete areas of the brainstem of
    suicide victims and control patients. Adv.Biochem.
    Psychopharmacol. <u>11</u>, 387-397.
Lloyd, K.G., Davidson, L. and Hornykiewicz, O. (1975).

The neurochemistry of Parkinson's disease: effect
of L-dopa therapy. J.Pharmacol. 195, 453-464.
Mackay, A.V.P., Yates, C.M., Wright, A., Hamilton,P.
and Davies, P. (1978). Regional distribution of
monoamines and their metabolites in the human brain.
J.Neurochem. 30, 841-848.
Moses, S.G. and Robins, E. (1975). Regional distribution of norepinephrine and dopamine in brains of
depressive suicides and alcoholic suicides. Psychopharmacology Commun. 1, 327-337.
Olszewski, J. and Baxter, D. (1954). Cytoarchitecture
of the Human Brain Stem. Lippincott, Philadelphia
and Montreal.
Palkovits, M., and Jacobowitz, D.M. (1974). Topographic atlas of catecholamine and acetylcholine-
esterase-containing neurons in the rat brain.II.
Hindbrain (mesencephalon, rhombencephalon). J.Comp.
Neurol.157, 29-42.
Pare, C.M.B., Yeung, D.P.H., Price, K. and Stacey,R.S.
(1969). 5-hydroxytryptamine, noradrenaline and
dopamine in brainstem, hypothalamus, and caudate
nucleus of controls and of patients committing sui-
cide by coal-gas poisoning. Lancet II, 133-135.
Poitras, D. and Parent, A. (1978). Atlas of the distribution of monoamine-containing nerve cell bodies
in the brain stem of the cat. J.Comp.Neurol.179,
699-717.
Riederer, P. and Wuketich, St. (1976). Time course of
nigrostriatal degeneration in Parkinson's disease.
J.Neural Transmission 38, 277-301.
Riley, H.A. (1960). An Atlas of the Basal Ganglia,
Brainstem and Spinal Cord. Hafner, New York.
Rinne, U.K. and Sonninen, V. (1973). Brain catecholamines and their metabolites in parkinsonian
patients. Arch.Neurol. 28, 107-110.
Roizen, M.F., Kopin, I.J., Thoa, N.B., Zivin, J., Muth,
E.A. and Jacobowitz, D.M. (1976). The effect of
two anesthetic agents on norepinephrine and dopamine
in discrete brain nuclei, fiber tracts, and terminal regions of the rat. Brain Res. 110, 515-522.
Sano, I., Gamo, T., Kakimoto, Y., Takesada, M. and
Nishinuma. K. (1959). Distribution of catechol
compounds in human brain. Biochim.Biophys.Acta 32,
586-587.
Shimada, S., Ishikawa, M. and Tanaka, C. (1976). Histochemical mapping of dopamine neurons and fiber
pathways in dog mesencephalon. J.Comp.Neurol.168,
533-543.
Spokes, E.G.S. and Koch, D.J. (1978). Postmortem
stability of dopamine, glutamate decarboxylase

23

and choline acetyltransferase in the mouse brain
under conditions simultating the handling of human
autopsy material. J.Neurochem. 31, 381-383.
St.Laurent, J., Roizen, M.F., Miliaressis, E., and
Jacobowitz, D.M. (1975). The effects of self-
stimulation of the catecholamine concentration of
discrete areas of the rat brain. Brain Res. 99,
194-200.
Tassin, J.P., Chermay, A., Blanc, G., Thierry, A.M.
and Glowinski, J. (1976). Topographic distribution
of dopaminergic innervation and of dopaminergic
receptors in the rat striatum. I. Microestimation
of (3H) dopamine uptake and dopamine content in
microdiscs. Brain Res. 107, 291-301.
Versteeg, D.H.G., Vanderguyten, J., DeJong, W. and
Palkovits, M. (1976). Regional concentrations of
noradrenaline and dopamine in rat brain. Brain
Res. 113, 563-574.
Winblad, B., Bucht, G., Gottfries, C.G. and Roos,B.E.
(1979) Monoamines and monoamine metabolites in
brains from demented schizophrenics. Acta psych.
scand. 60, 17-28.

Brain Neurotransmitter Amines in Cerebral Ischemia and Stroke

K. JELLINGER and P. RIEDERER

Ludwig Boltzmann-Institute of Clinical Neurobiology,
Lainz-Hospital, A-1130 Vienna, Wolkersbergenstraße 1,
Austria

SUMMARY

Disturbances of catecholamine metabolism and alterations of other putative neurotransmitters have been
demonstrated in both experimental cerebral ischemia
and in human stroke. In acute cerebral infarction
depletion of dopamine (DA), serotonin (5-HT) and
5-hydroxyindole acetic acid (5-HIAA) in both the
necrotic and intact brain areas is associated with
increase of the 5-HT precursor tryptophan (TRP) and
its major metabolite kynurenine (KYN), suggesting
increased release and decreased degradation of indoles
due to reduced activities of related oxygen dependent
enzymes, eg. tyrosine hydroxylase (TH). Specific
5-HT-binding to microsomal membrane fractions obtained
from necrotic brain tissue is significantly reduced,
but no decrease in binding capacity is found in intact
tissue of infarcted brain. In old cerebral infarction
reduction of 5-HT and 5-HIAA with slight increase of
TRP is seen in the necrotic area, while almost normal
levels of 5-HT, 5-HIAA, TRP and KYN in the surrounding
scar and intact brain tissue indicate "normalization"
of 5-HT and TRP metabolism in the previously ischemic
brain. There is experimental and human post-mortem
evidence that unilateral focal ischemia produces
bilateral effects on brain monoamines. Recent studies
in experimental and human cerebral ischemia and stroke
further indicate that disorders in the metabolism of
brain monoamines and other putative neurotransmitters
contribute to the development of postischemic brain
damage and the complicating cerebral edema.

25

INTRODUCTION

Disturbances in the function of brain monoamines and other putative neurotransmitters have recently assumed increasing importance among the metabolic derangements resulting from cerebral ischemia. Neurotransmitters are confined to neurons and not found in glia and vascular tissue. Thus alterations in neurotransmitter metabolism in ischemia and brain infarction reflect neuronal metabolism. Although most of the oxygen consumed by the brain is used to burn glucose and provide energy, a small amount is used in the synthesis of neurotransmitters. Norepinephrine (NE), and DA are synthesized from TYR and oxygen, 5-HT from TRP and oxygen. The enzyme catalyzing the first step of catecholamine (CA) synthesis is TH which is inhibited by CA and appears to be regulated in vivo by a still poorly understood activating system (Davis et al.,1979). While a number of mechanisms may limit the hydroxylation of TYR in the brain, one important factor appears to be the availibility of oxygen to CA neurons. The hydrolases which synthesize CAs and 5-HT use molecular oxygen, and their Km for oxygen is close to the physiological pO_2 in brain. It has been shown that brain CA and 5-HT synthesis are dependent on arterial oxygen level (Davis et al.,1979). Therefore, among the mechanisms that limit hydroxylation of TYR and TRP in the brain one important factor appears to be availibility of oxygen to CA and 5-HT neurons. In accordance with these properties, hypoxia and ischemia have been found to have a selective effect on the neurons that store CAs, 5-HT and other putative neurotransmitters. Experimental studies have shown that the synthesis and probably the metabolism of certain neurotransmitters are vulnerable to even mild hypoxia (Davis et al.,1979; Siesjö,1978) and ischemia (Moskowitz and Wurtman,1976; Mrsulja et al.,1978; Welch et al., 1977; Cvejic et al.,1980). Biosynthesis of amino acids that interact closely with putative neurotransmitters have also been found to be sensitive to hypoxia and ischemia (Siesjö,1978) and Yanagihara (1976) have reported that even brief episodes of anoxia can cause long-lasting impairment of cerebral protein synthesis which persists in vivo.
The depletion of brain monoamines, their release into the extraneuronal compartment, with subsequent accumulation of their metabolites and alterations of other neurotransmitter systems may contribute to a number of complications of ischemia including cerebral edema, and the progression of brain infarction

26

(Welch et al.,1977; Gaudet et al.,1978; Moskowitz and Wurtman,1976; Jellinger et al.,1978). Following cerebral reperfusion after periods of ischemia,dysfunction of neurotransmitters may persist despite improved cerebral energy metabolism and thus in a vicious circle may be responsible for post-ischemic deficit in neurological functions and permanent morphological damage (Gaudet et al., 1978; Zervas et al.,1975; Moskowitz and Wurtman,1976; Mrsulja,1979). Disorders of CA metabolism and alterations of other putative neurotransmitters, e.g. GABA and cyclic AMP, have been demonstrated in a variety of experimental models of cerebral hypoxia and ischemia (Lust et al.,1975; Welch et al.,1976; Kogure et al.,1976; Mrsulja et al., 1978; Flamm et al., 1978) as well as in both experimental and human cerebral infarction (Welch et al., 1975).

EXPERIMENTAL CEREBRAL ISCHEMIA

After acute induction of cerebral ischemia which results in a rapid depletion of cerebral energy reserves (Levy and Duffy,1977; Kobayashi et al.,1977; Siesjö,1978; Davis et al.,1979), cerebral monoamines undergo major changes due to inhibition of energy-dependent synthesis, degradation and transport mechanisms (Kogure et al.,1976; Lavyne et al., 1975; Moskowitz and Wurtman,1976; Welch et al.,1977; Mrsulja et al.,1978). CA brain levels have been evaluated at various times after induction of cerebral ischemia and brain infarction(for review see Jellinger et al.,1978; Gaudet et al.,1978; Bralet et al., 1979; Cvejic et al.,1980). Several studies indicate that there are no or only minimal changes in tissue levels of CA and 5-HT during short-term ischemia (Lust et al., 1975; Mrsulja et al.,1976; Bralet et al.,1979; Harrison et al.,1979), and that significant alterations in monoamine turnover occur in the postischemic states, i.e. during reperfusion after transient brain ischemia (Mrsulja et al., 1978; Gaudet et al.,1978; Calderini et al.,1978) suggesting a "maturation phenomenon" (Mrsulja, 1979). Recent data also show that alterations in monoamine metabolism persist for a long period following brief (15 min.) brain ischemia despite improved cerebral energy metabolism (Gaudet et al.,1978; Cvejic et al.,1980).

Among the putative neurotransmitters, NE, GABA, and cAMP are the first to be changed in ischemia (Kogure

27

et al.,1976; Mrsulja et al.,1978; Flamm et al.,1978),
while the release of DA seems to be more resistant
than that of other monoamines, particularly 5-HT,
the latter being highly sensitive to the effects of
even minimal ischemia (Gaudet et al.,1978; Mrsulja
et al.,1978). In the baboon, release of 5-HT from the
brain into cerebral arteriovenous blood has been
observed within one minute of occlusion of a major
cerebral artery (Welch et al.,1977). Ischemia produced
by unilateral cerebral embolism also causes rapid
depletion of 5-HT and NE but initially increases DA
in the rat cerebral cortex (Kogure et al.,1976).
After incomplete ischemia of 15 min. duration there
is increase in tissue levels of TYR, DA, TRP and
5-HIAA and pronounced decrease in NA and 5-HT with
increase in the rate of TYR hydroxylation and decrease
in TRP hydroxylation (Calderini et al.,1978). Decrease
of cAMP as early as 3 hours after the onset of
ischemia, when such lesions are reversible, indicate
that an initial step in the alteration of cellular
metabolism following ischemia is due to membrane
perturbation and altered production of this second
messenger (Flamm et al.,1978).

Long-term ischemia which is characterized by severe
reduction of tissue pO_2 and derangement of cerebral
energy state, is followed by suppression of cerebral
monoamine and protein biosynthesis, both being highly
energy-dependent processes. The results are severe
depletion of both CAs (DA and NE) and 5-HT (Brown et
al.,1975; Moskowitz and Wurtman, 1976; Kogure et al.,
1976; Welch et al.,1977; Mrsulja et al.,1978; Cvejic
et al.,1980; Gaudet et al.,1978), and disorders of
amino acid metabolism (Kobayashi et al.,1977;
Yanagihara,1978; Calderini et al.,1978). After
long-term ischemia brain levels of TRP, the natural
precursor of 5-HT, are significantly reduced, follcwed
by an increase during recirculation (Brown et al.,1975;
Ljunggren and Brown,1975). HVA and 5-HIAA, are not changed
initially, while in the post-ischemic phase both
these main metabolites of DA and 5-HT accumulate in
the damaged tissue and adjacent edema zone even when
inhibition of synthesis of their precursors was produced (Mrsulja et al.,1976,1977; Welch et al.,1976,1977).
Accumulation of monoamine metabolites and GABA in
ischemic brain has been related to (a) increased
synthesis or turnover and (b) decreased degradation
and washout (Mrsulja et al.,1978). On the other hand,
a marked increase of cerebral NA and DA levels was
observed in the early post-ischemic period after

28

transient cerebral ischemia produced by embolism
(Kogure et al.,1976; Ishihara et al.,1979) or vascular
occlusion (Mrsulja et al.,1976; Cvejic et al.,1980).

This has been termed "catecholamine rebound" and may
represent a state of post-ischemic hypermetabolism
(Gaudet et al.,1978). It seems unlikely, to be related
to reactive hyperemia or hypoxia in brain since tissue
pO_2 and regional CBF returned to steady state levels
when CA rebound was at its peak. An alternative explanation for increased DA and NE levels in ischemic
cerebral cortex could be persistent impairment of
degradation (Brown et al.,1975) coinciding with recovery of synthesis, since some degradative enzymes,
e.g. monoamine oxidase A and B and catechol-O-methyl
transferase (COMT) activities (Cvejic et al.,1980)
as well as TRP hydroxylase (Calderini et al.,1978)
have been found to be reduced in recent periods of
post-ischemia period (Mrsulja et al.,1978; Gaudet et
al.,1978; Ishihara et al.,1979) might stimulate CA
synthesis by feed-back mechanisms (Costa et al.,1974;
Lydiard et al.,1975). However, the nature of post -
ischemic rebound phenomena still requires to be
defined.

In addition, neither is there agreement on the pattern
of changes observed during ischemia of various duration, nor have consistent changes been reported in
post-ischemic states of various origin, e.g. complete
vs incomplete ischemia (Calderini et al.,1978), embolism vs vascular occlusion (Zervas et al.,1975; Kogure
et al.,1976; Ishihara et al.,1979; Bralet et al.,1979),
etc. There is evidence, therefore, that brain monoamine
changes may depend in part on the type of ischemia
(Siesjö,1978), the duration and on the method by which
ischemia is induced, the effects of embolism differing
from those of permanent vascular occlusion (Ishihara
et al.,1979) or between unilateral and bilateral
clamping of major cerebral arteries (Mrsulja et al.,
1976-1978; Cvejic et al.,1980).

Prolonged cerebral ischemia has been shown to cause
neurotransmitter disorders not only in the damaged
tissue but also in remote non-ischemic areas of the
brain, and in the contralateral hemisphere (Lavyne
et al.,1975; Welch et al.,1977; Lust et al., 1975;
Mrsulja et al.,1977,1979). Unilateral occlusion of
major cerebral arteries causing reduction of the
ipsilateral CBF (Nakai et al.,1977) results in depletion of 5-HT, DA and NE in the cerebral cortex of the

29

injured or of both hemispheres (Welch et al.,1977;
Gaudet et al.,1978; Mrsulja et al.,1978). Other   types
of unilateral cerebral ischemia caused by embolic
occlusion of the internal carotid artery (Bralet et
al.,1979) or the middle cerebral artery (Ishihara et
al.,1979) also cause DA and NA depletion on both cerebral hemispheres. The changes in the non-ischemic
hemisphere could either be a cause or an effect of
diaschisis, i. e. remote effect of a focal ischemia
coupled with reduced cerebral blood flow and edema
on cerebral metabolism and function (Levy and Duffy,
1977; Welch et al.,1977; Ishihara et al.,1979).

HUMAN CEREBRAL INFARCTION

Elevated levels of NA, 5-HT and cAMP have been measured
in the lumbar CSF of patients with cerebral infarct
and hemorrhage (Meyer et al.,1974; Welch et al.,1975),
while CSF HVA levels were significantly decreased in
patients with brainstem infarction (Hachinski et al.,
1978) and with spasms after aneurysm surgery the
latter also showing slight reduction of 5-HIAA and TRP
levels in ventricular CSF possibly related to cerebral
hypoxia (Vapahlahti et al.,1978). While these CSF
changes only provide limited information about cerebral monoamine metabolism in brain ischemia, recent
studies in human post-mortem brains largely confirmed
the data derived from experimental animals (Jellinger
et al.,1978; Riederer and Jellinger,1979).

A.RECENT BRAIN INFARCTION

In human brain severe to complete depletion of NE, DA
and 5-HT with significant reduction of its major metabolite 5-HIAA in the <u>necrotic area</u> of acute infarction
(table 1) is associated with considerable reduction
of both 5-HT and 5-HIAA in remote morphologically
intact areas of both hemispheres, the decrease of
both compounds being more pronounced in the damaged
hemisphere (table 2).
The perifocal <u>edema zone</u> shows almost normal DA levels,
with significant increase of both 5-HT and 5-HIAA
(table 1), while marked elevation of GABA, cyclic AMP,
TRP and its major metabolite KYN are seen in both the
necrotic area and perifocal edema zone. TRP shows
much higher elevation in the intact cortex and white
matter of the injured and less of the contralateral
hemispheres (table 3), while KYN displays three to
four-fold increase in the necrotic center of the infarction, and is much less elevated in the intact gray
and white matter (table 4).

TABLE 1. Post-Mortem Brain Concentrations of Monoamines in Recent Cerebral Infarction (Survival Time 6 - 8 hours)

Region examined		DA ng/g f.w. (n) mean ± S.E.M.	5-HT ng/g f.w. (n) mean ± S.E.M.	5-HIAA ng/g f.w. (n) mean ± S.E.M.
Necrotic area	cortex (I)	n.d. (11)	$2,3 \pm 0,9^{+)}$ (28)	$43,0 \pm 7,1^{+)}$ (16)
	cortex (H)	n.d. (3)	$19,7 \pm 1,2^{+)}$ (4)	n.e.
	white m.	n.d. (10)	$0,7 \pm 0,14^{+)}$ (27)	$23,9 \pm 8,8^{+)}$ (16)
	cp.striatum	n.e.	$3,0 \pm 1,5^{+)}$ (3)	$45,0 \pm 35,0^{+)}$ (3)
Perifocal edema	cortex	$26,1 \pm 6,95$ (7)	$81,8 \pm 18,7^{+)}$ (7)	$191,0 \pm 39,0^{+)}$ (9)
	white m.	$28,7 \pm 6,8$ (4)	$41,3 \pm 15,0$ (8)	$102,0 \pm 20,5^{+)}$ (10)
Intact cortex	ipsilat.	$31,0 \pm 2,55$ (5)	$18,7 \pm 10,9$ (4)	$69,5 \pm 1,5^{+)}$ (4)
	contralat.	$31,8 \pm 2,1$ (3)	$25,0 \pm 7,1$ (10)	$80,0 \pm 17,9$ (6)
Intact white m.	ipsilat.	n.e.	$2,5 \pm 1,4^{+)}$ (4)	$51,8 \pm 10,8^{+)}$ (4)
	contralat.	$40,0 \pm 2,0$ (2)	$15,0 \pm 6,0$ (6)	$50,0 \pm 19,0$ (5)
Controls	cortex	$25,0 \pm 5,0$ (9)	$55,0 \pm 7,0$ (9)	$119,0 \pm 8,0$ (9)
	white m.	n.e.	$25,0 \pm 4,2$ (7)	$47,0 \pm 4,1$ (7)
	cp. striatum	n.e.	$275,0 \pm 16,0$ (9)	$394,0 \pm 277,5$ (9)
	gl.pallidus	n.e.	$364,0 \pm 27,0$ (9)	$875,0 \pm 61,0$ (9)

+) = $p < 0,01$

n = number of assayed samples   I = ischemic necrosis   n.d. = not detectable

f.w. = fresh weight   H = hemorrhagic necrosis   n.e. = not examined

Specimens were obtained at autopsy. Post-mortem time ranged from 2 to 10 hours. DA, 5-HT and 5-HIAA were assayed fluorometrically using the methods of Anton and Sayre (1964), Ashcroft and Sharman (1962).

TABLE 2. Regional 5-HT and 5-HIAA in Intact Brain Areas in Human Cerebral Infarction

Brain Area		5-HT ng/g mean±S.E.M. Controls (n=9)	5-HT Acute	5-HT Old infarct	5-HIAA ng/g w.w. (mean ±S.E.M.) Controls (n=9)	5-HIAA Acute	5-HIAA Old infarct
Caudate n.	il	285±16	104 (2)	-	442±26	219 (2)	29 (1)
	cl	-	110 (2)	193 (2)	-	170 (2)	98 (1)
Putamen	il	266±16	197±55 (4)	109 (2)	535±28	571±144 (4)	475 (2)
	cl	-	107±39 (5)	145 (2)	-	476±128 (5)	-
G.pallidus	il	364±27	241±118 (4)	221±52 (3)	875±61	742±198 (4)	886 (2)
	cl	-	317 (2)	306±118 (3)	-	783±222 (3)	676±128 (3)
Thalamus	il	340±18	209±80 (6)	87±16 (4)	725±39	823±176 (5)	752 (2)
	cl	-	307 (2)	184±33 (4)	-	-	594±161 (3)
C.mamillare	il	250±18	158 (2)	226 (2)	-	-	-
	cl	-	496 (1)	496 (1)	-	996 (1)	-
Amygdal.n.	il	246±15	160±40 (4)	180 (2)	760±81	631±216 (3)	794 (2)
	cl	-	188 (2)	212 (2)	-	771	-
Hippocampus	il	200±11	72 (2)	212 (2)	480±21	387 (2)	767 (2)
	cl	-	-	-	-	-	-
N.ruber	il	603±29	441 (1)	186 (1)	1379±59	3358 (1)	2381 (1)
	cl	-	-	-	-	-	-
S.nigra	il	583±22	232±55 (3)	213±118 (3)	1595±47	1299±255 (3)	1113±328 (3)
	cl	-	-	120 (2)	-	-	1050 (2)
Form.ret.	il	476±25	350±95 (7)	1070±148 (3)	1035±42	2329±653 (7)	1273 (2)
	cl	-	41 (2)	30 (2)	-	-	-
Front.cort.	il	55± 7	38 (2)	30 (2)	119± 8	87 (2)	-
	cl	38	-	-	-	74 (2)	-

il = ipsilateral   n = number of cases examined
cl = contralateral   - = not examined

from: K.Jellinger et al. (1978)

TABLE 3. Post-Mortem Brain Concentration of TRP in Human Cerebral Infarction

Brain area	Cortex	White matter
Controls	$10,0 \pm 2,0$ (5)	$10,5 \pm 1,2$ (5)
Recent infarct (necrosis)	$15,0 \pm 1,2$[+)] (21)	$14,6 \pm 1,1$[+)] (16)
Perifocal edema	$16,8 \pm 4,0$[+)] (6)	$16,2 \pm 2,6$[+)] (7)
Intact ipsilat.hemisphere	$28,3 \pm 8,8$[+)] (3)	$21,6 \pm 7,7$[+)] (5)
Intact contralat. hemisph.	$21,0 \pm 7,2$[+)] (5)	$17,5 \pm 3,4$[+)] (6)
Old infarct (necrosis)	$12,8 \pm 0,5$ (4)	$11,5 \pm 1,2$ (3)
Perifocal Scar region	$11,3 \pm 1,2$ (5)	$11,8 \pm 2,4$ (4)

Values in µg/g w.w.   n = number of cases
mean ± S.E.M.   +) = $p < 0,01$
Post-mortem time ranged from 2 - 10 hours.
TRP was assayed using the method of Denckla and Dewey (1967)

TABLE 4. Post-mortem Brain Concentration of KYN in Human Cerebral Infarction

Brain area	Controls	Acute infarct Necrosis	Acute infarct Intact	Old infarct Necrosis	Old infarct Intact
Front.cortex	533 ± 86 (12)	1403 (2)	1750 (2)	1181 (2)	724 ± 231 (3)
Front.white m.	434 ± 49 (15)	1884 (2)	1120 (2)	393 (2)	565 ± 32 (3)
Caudate nucl.	542 ± 61 (14)	1632 (2)	1057 (2)	630 (2)	854 ± 295 (3)
Putamen	533 ± 76 (14)	1555 (2)	1217 (2)	543 (2)	845 ± 330 (3)
Amygdaloid n.	669 ± 106 (3)	2156 (2)	1220 (2)	1118 (2)	939 ± 130 (3)

Values in ng/g w.w.; mean ± S.E.M.;   n = number of cases
KYN was assayed by the method of Joseph et al. (1978) using
gas chromatography with electron capture detection.

Specific 5-HT binding is significantly decreased in
the necrotic area of acute brain infarct as compared
with the morphologically intact tissue of the
contralateral hemisphere (table 5). An almost identical 5-HT binding capacity is found in membrane fractions obtained from brains of patients without brain
dysfunction and from brain regions contralateral to
acute infarction.

TABLE 5. Specific 5-HT Binding in Human Cerebral
          Infarction
          (fmoles/mg protein); (number of patients)

Brain area	Controls	Intact tissue	Necrotic region
Frontal cortex	66,7 (1)	$68,2^{\pm}13,5$ (3)	$10,8^{\pm}3,2$ (2)
Frontal white m.	$98,3^{\pm}8,5$ (3)	$95,3^{\pm}15,6$ (2)	$16,4^{\pm}1,1$ (2)

For each region the specific 5-HT binding in two independently
isolated 45.000xg sediments from human post-mortem brain
homogenate were measured (method of Enna et al.,1977). All
binding experiments were performed six-fold (from Weiner et al.,
1979).

Tyrosine hydroxylase activity in post-mortem brain
tissue of two cases of vascular encephalopathy with
microinfarcts in the basal ganglia (age 72 and 76 years,
post-mortem time 2 and 4 hours) was significantly decreased in the basal ganglia, while the activity of
this enzyme in the brain stem was sligthly reduced
(table 6).

TABLE 6. Tyrosine Hydroxylase Activity in Human Post-
          Mortem Brain (nmoles DOPA/g/hours);
          Means $\pm$ S.E.M.

Brain area	Controls	Vascul.Encephalopathy
Caudate nucl.	$27,8 \pm 2,3$ (15)	3,35 (2)
Putamen	$16,2 \pm 5,9$ (5)	4,5 (2)
Subst.nigra	$19,4 \pm 6,2$ (4)	7,2 (2)
Hypothalamus	$3,1 \pm 1,0$ (5)	1,9 (2)
Corp.mamillare	$0,6 \pm 0,4$ (5)	0,4 (2)

Number of patients in parenthesis. The activity was measured
radioenzymatically using the method of McGeer et al. (1967).

34

Depletion of cerebral monoamines and increased release
of 5-HT in acute brain infarction associated with
increase of the 5-HT precursor TRP and its major
metabolite KYN could be explained by decreased
synthesis and turnover due to a lack of oxygen within
the necrotic area but also occur in remote regions
of the brain. The increase of 5-HT and TRP in the
surrounding edema region may be related either to
increased release with extracellular accumulation or
by decreased degradation and washout of these indoles
(Mrsulja et al.,1976,1978). The oxygen dependence of
TYR hydroxylase and TRP hydroxylase suggest that
ischemia should modify catecholamine and 5-HT synthesis in the brain (Davis et al.,1979). KYN, a major
metabolite of the 5-HT precursor TRP, synthesized by
the oxygen-dependent enzyme TRP-2,3-dioxygenase, has
also been shown to be present in and to be synthesized
within the brain (Joseph et al.,1978). More than 90 %
of TRP is converted via KYN and other metabolites to
nicotinamide. The oxygen-dependent enzyme TRP-2,3-di-
oxygenase is the first step in the KYN pathway of TRP
catabolism. KYN determination after TRP loading is
used as an indicator of the activity of this enzyme
occurring in the brain (Joseph et al.,1979). Cerebral
KYN concentration in acute cerebral infarct parallels
the changes of TRP in both the necrotic area and the
intact brain as well as of 5-HT in the perifocal edema
region. These data give evidence for a disturbed
metabolism of TRP - KYN pathway on the basis of reduced TRP-2,3-dioxygenase activity. Reduced activity of
other oxygen-dependent enzymes, e.g. TYR- and TRP
hydroxylase, in cerebral ischemia also suggests that
neuronal activity in DA-ergic and 5-HT-ergic systems
is depressed in the post-ischemic period (Ishihara et
al.,1979; Mrsulja et al.,1978; Siesjö,1978; Cvejic et
al.,1980). Depletion of cerebral energy reserves also
leads to reduced 5-HT uptake in presynaptic nerve
endings, as indicated by almost total depletion of
specific 5-HT binding in acute cerebral infarct. NE
uptake in synaptosome preparations obtained from
ischemic brain is also largely decreased, but relative
release is increased which was attributed to modification of the synaptosomal membrane properties in brain
ischemia (Mrsulja et al.,1976; Cevjic et al.,1980).
Whether the affinity or the number of 5-HT receptors
are decreased in ischemic brain remains to be eluci-
dated. On the other hand, despite bilateral changes
in energy metabolism and in concentration of brain
monoamines after unilateral focal ischemia (Lavyne et

al.,1975; Lust et al.,1975; Welch et al.,1977; Gaudet
et al.,1978; Mrsulja et al.,1978) suggesting remote
effects of ischemia coupled with reduced rCBF and
edema, no reduction of specific 5-HT binding was observed in morphologically intact areas of the brain
contralateral to the infarct.

Disorders of indoleamines and other putative neurotransmitter functions in ischemia and stroke may
be responsible for progression of post-ischemic brain
damage and the complicating edema (Zervas et al.,1975;
Moskowitz and Wurtman,1976; Jellinger et al.,1978;
Welch et al.,1977; Mrsulja,1979; Ishihara et al.,1979).
The increased release and turnover of 5-HT in the
perifocal edema zone of acute infarction is associated with marked elevation of GABA later returning to
normal values (Jellinger et al.,1978). These and other
biochemical changes,including transient alterations
of cAMP etc., may contribute to microcirculation disorders, impairment of collateral circulation and
progressive brain edema due to the experimentally
established effects of these substances on the
permeability of the blood brain-barrier (Welch et
al.,1977; Westergaard,1977). While the vasoconstric-
tive effect of 5-HT is debatable (Hardebo et al.,1978;
McKenzie et al.,1978), the changes of monoamines and
other putative neurotransmitters are clearly related
to cerebral edema and are suggested to contribute
to the development of post-ischemic tissue damage
and neurologic deficits. The relevance of the monoamine
changes to the progression of cerebral ischemia and
the complications associated with cerebral stroke,
however, remain to be established.

B. OLDER BRAIN INFARCTS

In infarction survived for more than 5 days a total
depletion of DA and 5-HT with reduction of 5-HIAA in
necrotic areas is associated with mild elevation of
DA and almost normal concentrations of 5-HT and
5-HIAA and normal values of GABA in the surrounding
degradation and scarring zone (table 7).

These data indicate "normalization" of the previously
increased 5-HT turnover after decrease of the acute
perifocal edema (Welch et al.,1977; Jellinger et al.,
1978). Only in the surrounding white matter there
is still a significant reduction of 5-HT with increased levels of 5-HIAA, while mild reduction of DA and

TABLE 7.   Post-Mortem Brain Concentrations of
           Monoamines in old Cerebral Infarction
           (Survival Time 5 Days - 9 Months)

Region examined	DA ng/n f.w. (n) mean±S.E.M.		5-HT ng/g f.w. (n) mean±S.E.M.		5-HIAA ng/n f.w. (n) mean±S.E.M.	
Necrotic area						
Cortex	n.d.	(8)	$0,35 \pm 0,35^{+}$	(14)	$66,0 \pm 23,0^{+)}$	(9)
White m.	n.d.	(8)	$1,4 \pm 1,3^{+)}$	(10)	$42,0 \pm 25,0_{+)}$	(5)
Cp. striatum	n.e.		$43,0 \pm 22,0^{+)}$	(3)	$102,0 \pm 54,0^{+)}$	(3)
Gl.pallidus	n.e.		$11,0 \pm 5,0^{+)}$	(2)	$352,5 \pm 98,5^{+)}$	(2)
Perifocal area						
Cortex	$38,6 \pm 186$	(5)	$46,0 \pm 18,0_{+)}$	(12)	$79,0 \pm 28,0$	(10)
White m.	n.e.		$1,1 \pm 0,1^{+)}$	(7)	$63,0 \pm 25,0$	(7)

n = number of assayed samples   n.d. = not detectable
f.w. = fresh weight   n.e. = not examined
+)   = p < 0,01
Post-mortem times, methods and control levels see table 1.

5-HT without considerable changes of 5-HIAA is observed in most morphologically intact areas of the infarcted brain, except for mild reduction of 5-HT and
5-HIAA in the cerebral cortex and caudate nucleus of
both hemispheres (table 2). TRP and KYN levels in
old infarcts and in the surrounding degradation and
scar regions are only slightly increased or almost
normal (tables 3 and 4).

These data indicate that in old infarcts reduction
of 5-HT and 5-HIAA with slight increase of TRP and
KYN are still present in the necrotic tissue, while
the adjacent scar region shows almost normal levels
of 5-HT, 5-HIAA, TRP and KYN indicating some recovery
process of 5-HT and TRP metabolism in the previously
ischemic brain.

In summary, the biochemical findings in human brain
infarction which confirm previous results of experimental cerebral ischemia and CSF findings in human

stroke strongly indicate that disorders of cerebral
metabolism of monoamines and other putative neurotransmitters and, in particular, of indoleamine
function, are contributing to the development of
post-ischemic brain damage and the complicating edema,
the decrease of which in the areas surrounding old
infarcts correlates well with some recovery process
of 5-HT metabolism in previously ischemic brain.
However, the pathogenesis and exact time course of
disorders of catecholamine metabolism and 5-HT receptor functions in cerebral ischemia and their significance in human stroke need further elucidation.

REFERENCES

Anton, A.H., Sayre, D.F. (1964). The distribution
of dopamine and dopa in various animals and a method
for their determination in diverse biological
materials. J.Pharm.exp.Ther. 145, 326-336.

Ashcroft, G.W., Sharman, F.D. (1962). Drug induced changes in the concentration of 5-OR indolyl
compounds in the cerebrospinal fluid and caudate
nucleus. Brit.J.Pharm.19, 153-160.

Brown, R.M., Kehr, W., and Carlsson, A. (1975).
Functional and biochemical aspects of catecholamine
metabolism in brain under hypoxia. Brain Res. 85,
491-509

Bralet, J., Beley, P., Bralet, A.M., Beley A.
(1979). Catecholamine levels and turnover during
brain ischemia in the rat. J.Neural Transm.48,143-155

Calderini C., Carlsson, A., Nordström, C.-H. (1978).
Influence of transient ischemia on monoamine metabolism in the rat during oxide and phenobarbitone
anesthesia. Brain Res. 157, 303-310

Cvejic, V., Micic, D.V., Djuricic, B.M., Mrsulja,
B.J., Mrsulja, B.B.(1980).Monoamines and related enzymes in
cerebral cortex and basal ganglia following transient
ischemia in gerbils. Acta neuropath.(Berl.) 51, 71-77

Costa, E., Guidotti, A., Živković, B. (1974).
Short- and long-term regulation of tyrosine hydroxylase. In Neuropsychopharmacology of Monoamines and
their Regulatory Enzymes,(ed. E.Usdin), pp 161-175,
Raven Press, New York.

Davis, J.N., Girson, L.T.,Jr., Stanton, E.,
Maury, W. (1979). The effect of hypoxia on brain
neurotransmitter systems. Advances in Neurology,
Vol 26, 219-223.

Denckla, W.D., Dewey, H.K. (1967). The determination of tryptophan in plasma, liver and urine.
J.Lab.clin.Med.69, 929-934.

Enna,S.J., Bennett, J.P.Jr., Byland,D.D., Creese,
I., Burt,D.B., Charness, M.D., Yamamura, H.I., Simatov,
R., Snyder,S.N. (1977). Neurotransmitter receptor
binding: Regional distribution in human brain.
J.Neurochem. 28., 233-236.

Flamm, E.S., Schiffer J., Viau, A.T., Naftchi,
N.E. (1978). Alterations of cyclic AMP in cerebral
ischemia. Stroke 9, 400-402.

Gaudet, R., Welch, K.M.A., Chabi, E., Wang, T.P.
(1978). Effect of transient ischemia on monoamine
levels in the cerebral cortex of gerbils. J.Neuro-
chem. 30, 751-757.

Hachinski, V., Shibuya, M., Norris, J.W., Hornykiewicz, O. (1978). Cerebrospinal fluid homovanillic
acid in cerebral infarction. J.Neural Transm. Suppl.
14, 45-50.

Hardebo, J.E., Edvinsson, L., Owman, Ch.,
Svendgaard, N.A. (1978). Potentiation and antagonism
of serotonin effects on intracranial and extracranial
vessels. Neurology (Minneap.) 28, 64-70.

Harrison, M.J.D., Marsden, C.D., Jenner, P.(1979).
Effect of experimental ischemia on neurotransmitter
amines in the gerbil brain. Stroke 10, 165-168.

Ishihara, N., Welch, K.M.A., Meyer, J.S., Chabi,
E., Naritomi, H., Wang, T.-P.F., Nell, J.H., Hsu,M.-C.,
Miyakawa Y. (1979). Influence of cerebral embolism on
brain monoamines. J.Neurol.Neurosurg. Psychiat.42,
847-853.

Jellinger, K., Riederer, P., Kothbauer, P. (1978). Changes of some putative neurotransmitters in human cerebral infarction. J.Neural Transm., Suppl.14,31-44.

Joseph, M.H., Baker, H.P., Lawson, A.M. (1978). Positive identification of kynurenine in rat and human brain. Biochem.Soc.Trans. 6, 123.

Kobayashi, M., Lust, W.D., and Passonneau, J.V. (1977). Concentrations of energy metabolites and cyclic nucleotides during and after bilateral ischemia in the gerbil cerebral cortex. J.Neurochem.29,53-59.

Kogure, K., Scheinberg, P., Kishikawa, H., Busto, H. (1976). The role of monoamines and cyclic AMP in ischemic brain edema. In Dynamics in brain edema, (eds. H.M.Pappius and W.Feindell), pp 203-214. Springer, Berlin-Heidelberg-New York.

Lavyne, M., Moskowitz M.A., Larin, F., Zervas, N., Wurtman, R.J. (1975). Brain ^{3}H-catecholamine metabolism in experimental cerebral ischemia. Neurology (Minneap.) 25, 483-485.

Levy, D.E., Duffy, T.E. (1977). Cerebral energy metabolism during transient ischemia and recovery in the gerbil. J.Neurochem. 28, 63-70.

Ljunggren, B., Brown, K.M. (1975). Monoamine metabolism in rat brain after increased intracranial pressure. In Intracranial pressure II (eds. N.Lundberg, U.Ponten, M.Brock), pp 189-194. Springer, New York-Heidelberg-Berlin.

Lust, W.D., Mrsulja, B.B., Mrsulja, B.J., Passonneau, J.V., Klatzo I. (1975). Putative neurotransmitters and cyclic nucleotides in prolonged ischemia of the cerebral cortex. Brain Res. 98, 394-399.

Lydiard, R.B., Fossom, L.H., Sparber, S.B. (1975). Postnatal evaluation of brain tyrosine hydroxylase activity, without concurrent increase in steady-state catecholamine levels, resulting from dl-alpha-methyl-paratyrosine administration during embryonic development. J.Pharmacol.exp.Therap. 194, 27-36.

McKenzie, E.T., Stewart, M., Young, A.R., Harper, A.M. (1978). Cerebrovascular actions of serotonin. In Advances in Neurology, Vol. 20, (eds. J.Cervos-Navarro et al.)pp 77-83, Raven Press, New York.

McGeer, E.G., Gibson, S., McGeer, P.L. (1967).
Some characteristics of brain tyrosine hydroxylase.
J.Lab.clin.Med.,69, 929-934.

Meyer, J.S., Okamoto, S., Shimaza, K. (1974).
Disordered neurotransmitter function demonstrated by
measurement of norepinephrine and 5-hydroxytryptamine
in CSF of patients with recent cerebral infarction.
Brain 97, 655-664.

Moskowitz,M.A., Wurtman, R.J. (1976). Acute stroke
and brain monoamines. In Cerebrovascular diseases
(ed. P.Scheinberg), pp 153-166. Raven Press, New York.

Mrsulja, B.B., Lust, W.D., Mrsulja, B.J.,
Passenneau, J.V. (1977). Effect of repeated cerebral
ischemia on metabolites and metabolic rate in gerbil
cortex. Brain Res. 119, 480-489.

Mrsulja, B.B., Mrsulja, B.J., Cvejic, V., Djuricic,
B.M., Rogad, L. (1978). Alterations of putative neurotransmitters and enzymes during ischemia in gerbil
cerebral cortex. J.Neural Transm. Suppl. 14, 23-30.

Mrsulja, B.B., Mrsulja, B.J., Spatz, M.,Ito, U.,
Walker, J.T.,Jr., Klatzo, I. (1976). Experimental
cerebral ischemia in mongolian gerbils. IV.Behaviour
of biogenic amines. Acta neuropath. (Berl.)36,1.8

Mrsulja, B.B. (1979). Some new aspects of the
pathochemistry of the postischemic period. In Pathophysiology of Cerebral Energy Metabolism (eds. Mrsulja,
B.B,, Lj.M.Rakic, I.Klatzo, M.Spatz), pp 47-59,
Plenum Press, New York - London.

Nakai, K., Welch, K.M.A., Meyer, J.S. (1977).
Critical cerebral blood flow after unilateral carotid
occlusion in the gerbil. J.Neurosurg.Psychiat. 40,
595-599.

Riederer, P., and Jellinger K.,(1979). Brain
monoamines in cerebral infarction and coma. In Pathophysiology of Cerebral Energy Metabolism (eds. B.B.
Mrsulja, Lj.M.Rakic, I.Klatzo, M.Spatz), pp 121-142,
Plenum Press, New York-London.

Siesjö, B.K. (1978). Brain energy metabolism and
catecholaminergic activity in hypoxia, hypercapnia
and ischemia. J.Neural Transm. Suppl.14, 17-22.

Vapahlahti, M., Hyyppä, M.T., Nieminen, V.,
Rinne, U.K. (1978). Brain monoamine metabolites and
tryptophan in ventricular CSF of patients with spasm
after aneurism surgery. J.Neurosurg. 48, 58-63.

Weiner, N., Martin, C., Wesemann, W., Riederer, P.
(1979). Effects of post-mortem storage on synaptic
5-HT binding and uptake in mammalian brain. J.Neural
Transm. 46, 253-262.

Welch, K.M.A., Chabi, E., Buckingham, J., Bergin,
B., Achar, V.S., Meyer, J.S. (1977). Catecholamine
and 5-HT levels in ischemic brain. Influence of
p-chlorophenylalanine.   Stroke 8, 341-346.

Welch, K.M.A., Chabi, E., Dodson, R.F., Bergin,B.
(1976). The role of biogenic amines in the progression
of cerebral ischemia and edema. In Dynamics of Brain
Edema (eds. Pappius, H., W. Feindel), pp 193-202,
Springer, Berlin-Heidelberg-New York.

Welch, K.M.A., Meyer, J.S., Chee, A.N.C. (1975).
Evidence for disordered cyclic AMP metabolism in
patients with cerebral infarction. Europ.Neurol. 13,
144-154.

Westergaard, E. (1977). The blood-brain-barrier
to horseradish peroxidase under normal and experimental conditions. Acta neuropath. (Berl.) 39, 181-186.

Yanagihara, T. (1976). Cerebral anoxia: effect on
neuron-glia fraction and polysomal protein synthesis.
J.Neurochem. 27,537-544.

Zervas, N.T., Lavyne,M.H., Negoro, M. (1975). Neurotransmitters and the normal and ischemic cerebral
circulation.   New Engl.J.Med.293, 812-816.

[^{18}F]Fluoro-Dopa for the *in vivo* Investigation of Intracerebral Dopamine Metabolism

E. S. GARNETT, G. FIRNAU and C. NAHMIAS

McMaster University Medical Centre,
Hamilton, Ontario, Canada

Although a deficiency of dopamine in the nigrostriatal tracts is known to be associated with Parkinson's disease (Hornykiewicz, 1966) and the dopamine hypothesis of schizophrenia has been extant for some years, there is still no method by which direct measurements of intracerebral dopamine metabolism can be made during life. Indirect assessments of intracerebral catecholamine metabolism, based on examination of the cerebrospinal fluid have been reported (Van Praag and Korf, 1975). However the cephalad to caudad concentration gradient of amine metabolites in the cerebrospinal fluid and the need to block their reabsorption by the weak organic acid probenecid renders such measurements of limited value. Attempts to measure intracerebral dopamine metabolism based on a knowledge of the veno arterial concentration difference across the brain have not been successful because the amount of metabolites released from the brain are small (Fyro, Settergren and Sedvall, 1975). In our laboratory we have taken a different approach to the problem. We have measured the rate of change of concentration of ^{18}F-fluorine in the brain after an intravenous injection of [^{18}F]5-fluoro dopa and have derived values for intracerebral dopamine metabolism from this information.

3-4-Di-hydroxy-phenyl-alanine, dopa, the immediate precursor of dopamine was labelled with the positron emitting isotope fluorine-18 by a modification of the Balz Schiemann reaction (Firnau, Nahmias and Garnett, 1973). ^{18}F was chosen as the label because it has a physical half life long enough for chemical syntheses (110 min). It was also argued that the small ionic radius of fluorine would not influence the biological behaviour of dopa if the fluorine atom was introduced into the 5 position on the benzene ring.

Having synthesised [18F]fluoro dopa it was first necessary to show that its biological behaviour is indistinguishable from that of native dopa. The rate of clearance of [18F]fluoro dopa from the blood and its organ distribution in rabbit and rat is the same as that of [14C] dopa (Garnett and Firnau, 1973). The brain uptake index of [18F]-fluoro dopa is also the same as that of [14C] dopa (Garnett, Firnau, Nahmias et al, 1980). When fluoro dopa is the substrate for dopa decarboxylase (EC 4.1.1.26) derived from hog brain or hog kidney cortex, the Km of the reaction is similar to that obtained when dopa is the substrate; Km fluoro dopa ranged from 1.68 to 2.31 mmol/l, Km dopa 3.0 mmol/l (Firnau, Garnett, Sourkes et al, 1975). Like dopa, fluoro dopa is methylated by catechol-O-methyl-transferase (EC 2.1.1.6). The major site of methylation is however the 4-hydroxyl group rather than the 3-hydroxyl group. This difference we attribute to the ionisation of the 4-hydroxyl group that results from the juxta position of the fluorine atom in 5-fluoro dopa (Firnau, Pantel and Garnett, 1980). 5-Fluoro dopamine binds to dopamine receptors derived from a homogenate of rat caudate nucleus and competes with [3H]-spiroperidol. The C_{50} for this competition is 63×10^{-6}M. 5-Fluoro dopamine also stimulates the release of cyclic-AMP by adenylate cyclase. At physiological pH the efficiency with which it does this is about half of that of dopamine. The pK_{OH} of 5-fluoro dopamine is 8.0 compared to 8.9 for dopamine. It is believed that both hydroxyl groups of dopamine should be unionised for maximal interaction at the receptor. Given the difference in the pK_{OH} of the molecules it seems likely that the reduced stimulation of adenylate cyclase at pH 7.4 results from ionisation of the 4-hydroxyl group of 5-fluoro dopamine.

On the basis of the above results the similarity between fluoro dopa and dopa was considered such that the [18F]5-fluoro dopa could be used as a tracer for intracerebral dopamine metabolism.

[18F]5-Fluoro dopa was injected intravenously into female cynomolgus monkeys anaesthetised with enflurane and the rate of appearance of [18F] in the brain and the rate of disappearance of [18F] from the blood was measured. It was then possible to calculate fractional rate constants for the transfer of dopa into the brain and its subsequent metabolism within the brain. A three compartment explanatory model was used and corrections for [18F] contained in the brain were made using [113mIn]-indium transferin as a blood pool marker. The details and justification for the model have recently been described (Garnett, Firnau, Nahmias et al, 1980). The fractional rate constants derived from the model were as follows: forward flux of L-dopa into the endothelium of the brain capillaries from the blood (1.54 ± 0.37 min^{-1}), back flux of dopa and/or its metabolites to the blood from the endothelium (0.67 ± 0.05 min^{-1}), formation of dopamine from dopa

(0.20 $\pm$ 0.04 min^{-1}) and destruction of dopamine (0.04 $\pm$ 0.02 min^{-1}). Using these values of the fractional rate constants and the known concentration of dopa in the blood and dopamine in the brain we were able to show that the absolute rates of transfer of L-dopa across the blood brain barrier, and the turnover of dopamine within the brain, were very similar to those reported from experiments that had necessitated decapitation of the animal. Of particular interest, it was shown that only one fifth of the dopa that entered the endothelium was available for dopamine synthesis in neurones. Preliminary experiments in rats suggest that this figure may be as low as 10% in rodents.

In the course of the work on cynomolgus monkeys it was also shown that the rate of appearance of DL[^{18}F]O-methyl fluoro-dopa was slow. Thus it could reasonably be assumed that only fluoro dopa was present in the blood feeding the brain; provided the experiments were confined to the 10 min immediately after the injection of the tracer.

In another series of experiments done on conscious female baboons (Papio papio), it was possible to demonstrate changes in the pattern of accumulation of ^{18}F in the brain after pharmacological manipulation with α-methyl dopa, reserpine, pargyline and haloperidol (Garnett, Firnau, Chan et al, 1978). In control animals into which [^{18}F]fluoro dopa had been injected intravenously ^{18}F accumulated continuously in the head. α-Methyl dopa, a competitive inhibitor of dopa transport and decarboxylation prevented the accumulation of ^{18}F; reserpine, known to release stored intracerebral dopamine, discharged ^{18}F; pargyline, a mono amine oxidase inhibitor, and haloperidol, a known augmentor of intracerebral dopamine turnover, increased the rate of accumulation of ^{18}F. These changes in the accumulation of intracerebral ^{18}F were commensurate with the known action of the drugs used to induce them. Thus it was confirmed that intracerebral dopamine metabolism could be monitored externally and atraumatically in fully conscious primates.

We conclude that fluorine-18 labelled precursors of central amines can be used to obtain quantitative data on the intracerebral metabolism of these molecules. Dopamine, noradrenaline and 5-hydroxy tryptamine can all be studied in the ways discussed in this paper.

With the increasing availability of positron emission tomography it will soon be possible to make regional as well as global measurements of intracerebral amine metabolism and so derive new insight into the role of the amines and similar agents in the control of mood and locomotion.

ACKNOWLEDGEMENT

The authors acknowledge the financial assistance of the
Medical Research Council and the Ontario Mental Health Foundation
Canada.

REFERENCES

Firnau, G., Garnett, E.S., Sourkes, T.L. and Missala, K.(1975).
[^{18}F]Fluoro-dopa: a unique gamma emitting substrate for dopa
decarboxylase. Experientia, 31, 1254-1255.

Firnau, G., Nahmias, C. and Garnett, E.S.(1973). The preparation
of [^{18}F]5-fluoro-dopa with reactor produced fluorine-18. Int.J.
Appl.Radiat.Isot. 24, 181-184.

Firnau, G., Pantel, R. and Garnett, E.S.(1980). Ionisation of
the catechol in dopa changes the site of methylation by catechol-
O-methyl transferase. Molec. Pharmacol. (in press).

Fyro, B., Settergren, G. and Sedvall, G.(1975). Release of homovanillic acid from the brain of children. Life Sci. 17, 397-402.

Garnett, E.S. and Firnau, G.(1973). [^{18}F]5-Fluoro-dopa as a new
brain scanning agent. In Radiopharmaceuticals and Labelled
Compounds. IAEA-SM-171/77, 1, Vienna.

Garnett, E.S., Firnau, G., Chan, P.K.H., Sood, S. and Belbeck,
L.W. (1978). [^{18}F]Fluoro-dopa, an analogue of dopa, and its use
in direct external measurements of storage, degradation, and
turnover of intracerebral dopamine. Proc. Natl. Acad.Sci.
(USA), 75, 464-467.

Garnett, E.S., Firnau, G., Nahmias, C., Sood, S. and Belbeck, L.W.
(1980). Blood brain barrier transport and cerebral utilization
of dopa in living monkeys. Am. J. Physiol., 238, R318-R327.

Hornykiewicz, O.(1966). Dopamine (3-hydroxy tyramine) and brain
function. Pharmac. Rev., 18, 925-964.

Van Praag, H.M. and Korf, J.(1975). The Dopamine hypothesis of
schizophrenia. In On the Origin of Schizophrenic Psychoses,
(ed. H.M. Van Praag), De Erven Bohn BV, Amsterdam.

Analytical Problems in Post-mortem Brain Studies

C. G. GOTTFRIES*, ROLF ADOLFSSON and BENGT WINBLAD

*Professor of Psychiatry at the University of
Göteborg, Department of Psychiatry and Neurochemistry,
St Jörgen's Hospital, S-422 03 Hisings Backa, Sweden*

Only in exceptional cases biopsies can be taken from the human brain. Direct investigations therefore have to be done at autopsy. Since long it is known that histological changes take place as well in normal ageing as in degenerative brain diseases. During the last two decades also neurochemical changes in the brain have been studied post-mortem and valid information has been achieved. At present there are a few centers in the world which are involved in a more systematical investigation of post-mortem human brain material. The methodology used at these centers is not uniform.

In this paper the following problems are discussed which we think are of importance at least when handling post-mortem brain material in the way we do:

Dissection technique
Circadian variations
Seasonal variations
Time death - autopsy
Sex
Age
Cause of death
Medication in terminal phase

Dissection technique
The brain is dissected immediately after it is taken out from the skull. The nucleus and brain regions which are of interest are deepfrozen in airtight plastic bags to -80 degrees of centigrade. The dissection of the brain is made after macroscopic anatomical landmarks. When obtaining tissue samples as much as possible of one nucleus is taken without including surrounding tissue. The reason for this is that we have found that one and the same nucleus in different parts may have different concen-

47

-trations of chemical substances. In 18 cases samples were removed from the anterior and posterior parts of the nucleus caudatus and the putamen and from the apical and the basal part of
the globus pallidus. In the caudate nucleus and in the putamen
the anterior parts contained significantly more homovanillic
acid (HVA) than the posterior parts. In the putamen there were
significantly higher concentrations of dopamine (DA) in the posterior part than in the anterior part and in the globus pallidus
the basal region contained significantly more HVA and DA than
the apical region (Adolfsson et al., 1979). We have also found
that those parts of the cortex cerebralis that belong to the
phylogenetically older regions, the limbic structures, have
significantly higher amounts of 5-hydroxytryptamine (5-HT) and
5-hydroxyindoleacetic acid (5-HIAA) than regions from the neocortex (Gottfries et al., 1974).

In 1975 Aquilonius et al. presented an easier and more accurate
method when dissecting post-mortem material. With the aid of a
cryomicrotome such as the LKB 2250 PMV Cryo-Microtome, exact
localization of areas of interest from a frozen human brain can
be obtained. Thin slices of the brain are cut from which samples are punched out with a cold micro pipette. The subnuclear
differentiation of the analyzed substances in an apparently homogenous structure can by this method better be guaranteed.

When more than one chemical analysis is to be made from one
brain sample the variation of the concentration in one and the
same nucleus must be considered. It is not possible to thaw the
deep frozen brain sample to get a homogenate which again is deep-
frozen. This procedure may cause degradation of chemical substances in the sample. Because of this we have homogenized the
brain sample by crushing the deepfrozen material to a powder
and in this way we get a homogenate which can be divided for the
different laboratory investigations. Much care must here be
taken so that for instance cold instruments always are used not
to thaw parts of the material.

Circadian Variations
Circadian variations of the levels of catecholamines and 5-HT
in the brains of experimental animals have been reported. In an
investigation by Carlsson et al. (1980) the hypothalamic nor-
adrenaline- (NA) level fluctuated significantly during the day
(Figure 1). The normetanephrine- (NM) and the 3-methoxy-4-
hydroxy-phenylenglycol- (MHPG) level showed similar variations.
There is a peak in the night in hypothalamus of NA and a nadir
between 9 and 12 in the day. Of interest is that the circadian
rhythm of NA almost mirror the circadian rhythm of cortisol
(Sachar, 1977). The level of DA and 5-HT in the hypothalamus
showed significant circadian variation with low levels around
6 - 9 a.m. The investigation by Carlsson et al. (1980) was made
on post-mortem human brain material where the hour of death was
recorded.

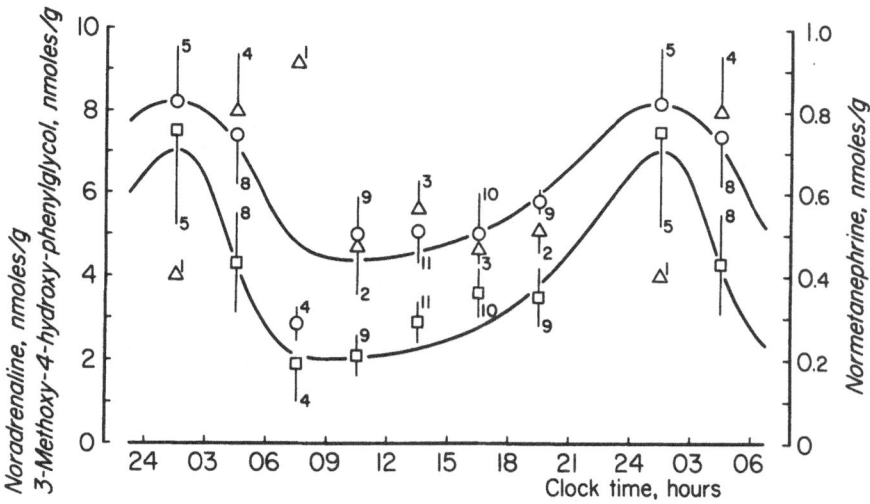

Figure 1. Levels of noradrenaline (O), 3-methoxy-4-hydroxy-phenylglycol (MHPG △) and normetanephrine (□) in human hypothalamus examined post mortem, in relations to clock time of death. Shown are the means ± s.e.m. (n) of pooled values of seven three-hour intervals. Statistics: (oneway analysis of variance followed by Student's t-test).

Seasonal Variations

In the paper by Carlsson et al. (1980) it was also shown that there was a seasonal variation in the level of 5-HT in the hypothalamus. A maximum was found during October – November and a minimum during December – January (Figure 2). No corresponding seasonal variation in hypothalamic 5-HIAA-levels were found. The hypothalamic DA also showed seasonal variation with possibly two peaks during January – February and August – September and two nadirs during March – June and October – December. In other brain regions the influence of biorhythms was less striking.

Time Death – Autopsy

Time elapsed between the occurence of death and autopsy is an essential variable due to the post-mortem degradation of both enzymes, monoamines and metabolites. From experience with animals (Wiesel and Sedvall, 1974) it has been shown that DA-levels are reduced in decapitated mice by 27 % after 6 hours and by 75 % after 48 hours at +4 degrees of centigrade. The degradation is temperature dependent and occurs mainly during the first few hours (Carlsson et al., 1974). According to the findings by Carlsson et al. it can be assumed that DA is metabolized post-mortem by catechol-0-methyltransferase to 3-methoxytyramine which process needs no oxygen. This metabolism certainly takes part immediately after the patient has died. In human studies

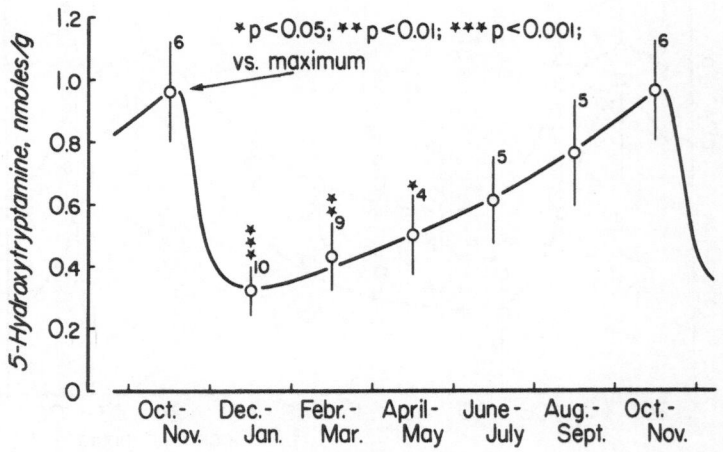

Figure 2. Level of 5-HT in human hypothalamus examined post mortem, in relation to month of death. Shown are the means ± s.e.m. (n) of pooled values of two consecutive months. Statistics: Student's t-test.

negative correlations between the DA-levels and the time elapsed between death and autopsy was found (Carlsson and Winblad, 1976). In a recently investigated material of 14 control cases this time variable was correlated to the chemical variables investigated (Table). The time between the occurrence of death and autopsy varied between 10 and 90 hours. As is evident from the table there is a significant negative correlation between this time variable and NA and NM in the hypothalamus, metatyrosine (MT), monoamine oxidase A (MAO-A) and cholinacetyltransferase (CAT) in the caudatus, and CAT in the cortex gyrus cinguli. A significant positive correlation was seen between the levels of the amino acids, tryptophane and tyrosine in all investigated areas and the time elapsed between the occurrence of death and autopsy (Table, Figure 3, 4). In previous investigations we have found that 5-HIAA seems to be sensitive for post-mortem degradation if long time has elapsed between the occurrence of death and autopsy (Beskow et al., 1976).

Sex
There are now more and more data supporting sex differences in the levels of monoamines, monoamine metabolites and related enzymes. We have found significantly higher 5-HIAA levels in hypothalamus, thalamus and cortex from lobus occipitalis and hippo-

50

	Hypo-thalamus	Caudatus	Hippo-campus	Cortex gyrus cinguli
5-HT	-0.48^{+}	-0.12	0.15	0.45
5-HIAA	-0.23	0.16	0.53^{+}	0.44
NA	-0.71^{**}	0.25	-0.04	-0.11
MHPG	-0.45	-0.01	-0.12	0.07
NM	-0.59^{*}			
DA	-0.12	-0.49^{+}		
HVA	-0.39	-0.41	0.51	0.17
MT		-0.81^{***}		
DOPA	0.25	0.09		
Tryptophane	0.62^{*}	0.60^{*}	0.66^{*}	0.70^{**}
Tyrosine	0.68^{**}	0.66^{*}	0.48^{+}	0.73^{**}
MAO-A	-0.47	-0.60^{*}	-0.32	-0.07
MAO-B	-0.06	-0.29	-0.40	0.16
CAT	-0.42	-0.78^{**}	-0.18	-0.53^{*}

Statistical analyses of correlations between time elapsed between the occurrence of death and autopsy. Figures are product-moment correlation coefficients.

+   = $p < 0.10$
*   = $p < 0.05$
**   = $p < 0.01$
***   = $p < 0.001$

-campus in females compared to males. In mesencephalon and pons the women also had higher levels of 5-HT (Beskow et al., 1976, Gottfries et al., 1974). The HVA-levels in the putamen were also higher in the female group compared to the male group.

Age
Age must continuously be under control when studying monoamines, their metabolites and enzyme activities in post-mortem investigations. For variation in age the presentation by Winblad et al. (1980) in this symposium is referred to,

Cause of Death

It can be assumed that patients with cancer who die in a very cachectic status, may have changed levels of monoamines, their metabolites or changed activity in enzyme systems. We have studied the levels of DA and HVA in a cancer group and in a non-cancer group. No significant differences were however found (Adolfsson et al., 1979). In some regions of the brain from the cancer group we have found higher 5-HIAA-levels. The high 5-HIAA-levels could not be referred to special cancer forms. It could be assumed that treatment with morphine before death could account for the high values of 5-HIAA in the cancer group. We found, however, no indication that the medication before death explained the differences (Gottfries et al., 1974).

Animal studies (Brown et al., 1975) have shown that hypoxia decreases the DA-turnover. Davis and Carlsson (1973) found that hypoxia has a strong and rapid effect on the hydroxylation of tyrosine and tryptophane and causes a decrease not only of DA but also of 5-HT and 5-HIAA. In patients who have suffered serious hypoxia before death we found reduced 5-HT and 5-HIAA-levels in the brain post-mortem (Gottfries et al., 1974).

Medication in Terminal Phase

It can be assumed that the kind of medical treatment the patient has got the last days of life may influence e.g. monoamines and their metabolites. So far we have, however, not found any serious influence of this kind. Patients treated with morphine or neuroleptics in conventional doses do not show any significant differences from patients not treated with this drugs. However, drugs with more direct effects on the metabolism of monoamines as MAO-inhibitors have an effect on the levels of the chemical substances (Ganrot et al., 1962). Of course the influence of drugs of precursors as L-dopa and tryptophane, given in the terminal phase, must continuously be checked when studying post-mortem brain material.

ACKNOWLEDGMENT

This investigation was supported by grants from Hjalmar Svensson's Foundation, Fredrik and Ingrid Thuring's Foundation and from the Swedish Medical Research Council (grant No. B81-05002-05).

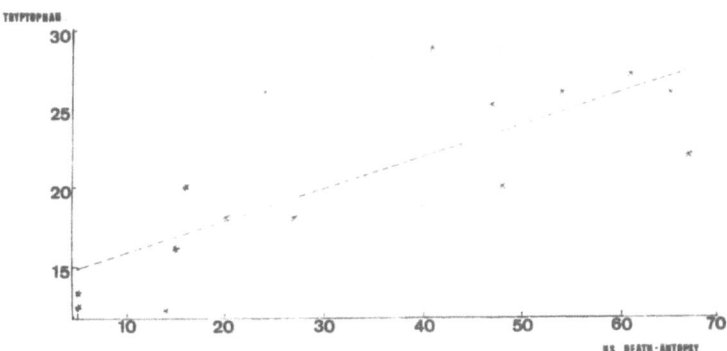

Figure 3. Increase of the levels of tryptophane in hippocampus according to the time elapsed between the occurrence
of death and autopsy.

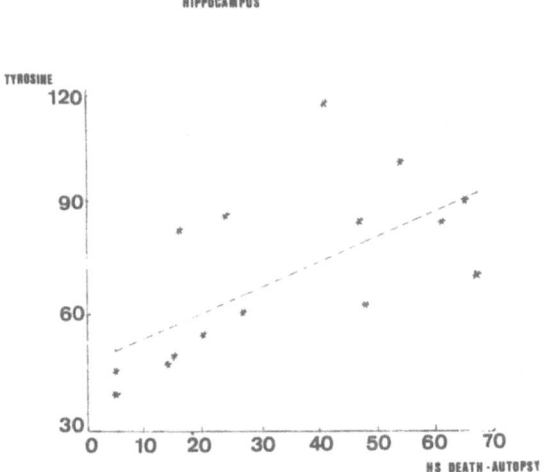

Figure 4. Increase of the levels of tyrosine in hippocampus
according to the time elapsed between the occurrence of death
and autopsy.

REFERENCES

Adolfsson, R., Gottfries, C.G., Roos, B.E. and Winblad, B. (1979). Post-mortem distribution of dopamine and homovanillic acid in human brain, variations related to age, and a review of the literature. J. Neural Transm., 45, 81 - 105.

Aquilonius, S.M., Eckernäs, S.A. and Sundwall, A. (1975). Regional distribution of choline acetyltransferase in the human brain: changes in Huntington's chorea. J. Neurol. Neurosurg. Psychiatry, 37, 669.

Beskow, J., Gottfries, C.G., Roos, B.E. and Winblad, B. (1976). Determination of monoamine and monoamine metabolites in the human brain: post-mortem studies in a group of suicides and in a control group. Acta Psychiatr. Scand., 53, 7 - 20.

Brown, R.M., Kehr, W. and Carlsson, A. (1975). Functional and biochemical aspects of catecholamine metabolism in brain under hypoxia. Brain Res., 85, 491.

Carlsson, A., Lindqvist, M. and Kehr, W. (1974). Postmortal accumulation of 3-methoxytyramine in brain. Naunyn-Schmiedebergs Arch. Pharmacol., 284, 365.

Carlsson, A., Svennerholm, L. and Winblad, B. (1980). Seasonal and circadian monoamine variations in human brains examined post mortem. In Biogenic amines and affective disorders. Proceedings of a symposium held in London, 18-21 January 1979, (eds. T.H. Svensson and A. Carlsson), Acta Psychiatr. Scand., 61, Suppl. 280.

Carlsson, A. and Winblad, B. (1976). Influence of age and time interval between death and autopsy on dopamine and 3-methoxytyramine levels in human basal ganglia. J. Neural Transm., 38, 271 - 276.

Davis, J.N. and Carlsson, A. (1973). The effect of hypoxia on monoamine synthesis levels and metabolism in rat brain. J. Neurochem., 21, 913 - 915.

Ganrot, P.O., Rosengren, E. and Gottfries, C.G. (1962). Effect of iproniazid on monoamines and monoamine oxidase in human brain. Experientia, 18, 1 - 5.

Gottfries, C.G., Roos, B.E. and Winblad, B. (1974). Determination of 5-hydroxytryptamine, 5-hydroxyindoleacetic acid and homovanillic acid in brain tissue from an autopsy material. Acta Psychiatr. Scand., 50, 496 - 507.

Sachar, E.J. (1977). Endocrine research in psychiatry. In Neurotransmission and disturbed behaviour, (eds. H.M. van Praag and J. Bruinvels), Bohn, Scheltema and Holkema, Utrecht.

Wiesel, F.A. and Sedvall, G. (1974). Post-mortal changes of dopamine and homovanillic acid levels in rat striatum as measured by mass fragmentography. Brain Res., 65, 547 - 550.

Regulation of 5-HT Receptor Binding as a Biochemical model for Pathological or Functional Changes in the CNS

WOLFGANG WESEMANN, NORBERT AROLD,
APPLETREE RODDEN and NINA WEINER
*Abteilung für Neurochemie, Physiologisch-Chemisches
Institut II, Philipps-Universität, Marburg/Lahn, West
Germany*

Chemically mediated neurotransmission and the transmitter-receptor interaction depend on the amount of
transmitter molecules released as quantal units into
the synaptic cleft [Del Castillo and Katz, 1955].
Aside from transmitter concentration also the quality
and quantity of receptor molecules will affect the
transmission process. Though biochemical in vitro
studies are restricted to the characterization of
ligand-receptor binding which can supply only
limited information about the possible physiological
response, the measurement of binding kinetics has
proved to be a suitable tool for describing central
dysfunctions on the molecular level. After cerebral
insult a significant decrease in 5-HT binding was
observed in membranes isolated from the infarct region as compared with fractions obtained from intact
regions or from control postmortem brain [Weiner et
al., 1979]. The possibility that changes in ligand-
receptor interaction reflect merely generalized
necrosis or postmortem changes seems unlikely in the
light of studies on 5-HT binding in Huntington's
chorea [Enna et al., 1976]. Here, the ligand binding
was found to be decreased specifically as far as
ligand and brain area are concerned. The decreased
5-HT binding in the caudate nucleus of choreatic
brain is caused by a reduction in the number of re-

ceptor sites and not by a decreased affinity towards
5-HT. The reduction in the number of receptor sites
may be either the consequence of a selective histopathological alteration in a specific brain area or
part of a regulation system which reduces the receptor sites in the presence of increased transmitter
concentrations. Both effects may be present in Huntington's chorea since both, neuropathological
changes [Stahl and Swanson, 1974] as well as increased 5-HT concentrations, had been observed
[Bernheimer and Hornykiewicz, 1973; Birkmayer, 1969].
The finding that after treatment of rats with MAO
inhibitors, which increase brain 5-HT concentration,
the number of receptors is decreased, supports the
idea that receptor cells have a control mechanism
which adapts the receptor sensitivity to the supply
of neurotransmitter [Savage et al., 1980]. In a few
pilot studies the "receptor sensitivity modification" has been used as a potential treatment of
tardive dyskinesia, schizophrenia, and Gilles de la
Tourette syndrome. In these cases patients showed a
marked improvement after prolonged treatment with
high doses of L-DOPA which is explained by a decrease
in the sensitivity of dopamine receptors persisting
even after subsequent withdrawal of the drug [Fried-
hoff and Alpert, 1978].

The question arises as to whether the pharmacological terms "supersensitivity" and "subsensitivity"
can be explained on the basis of high and low numbers of binding sites alone. In order to investigate
other possible factors regulating or modulating
transmitter-receptor interaction, studies on the
central 5-HT receptors have been performed which
may help in the understanding of functional or
pathological changes in neurotransmission as well as
preclude misinterpretation of in vitro results.

Effect of Buffer Composition on Ligand Binding Assay
Though the protein nature of receptors is indispu-
table the implication of this property has not always been well observed in ligand-binding assays.

TABLE 1. Effect of Ionic Strength on $[^3H]$5-HT Binding to Synaptic Membranes, P_3, of the Rat

Tris-HCl pH 7.2	Specific Binding *		n
	%	S_E	
1 mM	12.8 [+]	±1.7	12
25 mM	32.5	±1.3	16
50 mM	30.8	±3.8	12

* The specific binding is given as % of total binding. In each experiment 6 parallel incubations with 9 nM $[^3H]$5-HT were performed.

[+] $p \leqq 0.001$ different from 25 and 50 mM

57

TABLE 2. Effect of 0.1 % Ascorbate and 10 μM Pargyline on [^{3}H]5-HT Binding to a Crude Membrane Fraction Isolated from Postmortem Forebrain

Buffer Addition	Total Binding [fmoles/mg protein]	Specific Binding [fmoles/mg protein]	[% of total binding]
-	873	294	34
0.1% Ascorbate	901	180	20
10 μM Pargyline	1415	561	40
0.1% Ascorbate + 10 μM Pargyline	871	279	32

Homogenates from human forebrain were sedimented at 2 x 45,000 x g [Enna et al., 1977] and incubated in 1 mM Tris-HCl buffer (+ 1200 μg protein/ml) with 20.9 nM [^{3}H]5-HT in the presence or absence of pargyline and ascorbate. The binding assay was performed as described in the legend of Table 1 with 6 incubations, run in parallel.

58

The activity of an enzyme as well as the sensitivity of a receptor can be influenced not only by pH but also by the ionic or lipid environment. Table 1 shows that the specific 5-HT binding to a microsomal fraction of synaptic membranes, P_3 [Snyder and Bennett, 1975], is significantly enhanced if the molarity of the Tris-HCl buffer is increased from 1 mM up to 25 or 50 mM. This effect, though trivial, is sometimes neglected if the percentage of specific binding from total binding is taken as a parameter to compare the purity of different preparations.

A widely used buffer system in the radioligand-binding assay of catecholamines and 5-HT contains 0.1% ascorbate as antioxidant and 10 μM pargyline as MAO inhibitor [Snyder and Bennett, 1975]. However, ascorbate may affect the lipophilic environment of the receptor molecules within the membrane since it catalyzes the formation of lipid peroxides [Schaefer et al., 1975]. Accordingly, a reduction in the number of binding sites was observed for the opiate, the dopamine, the α-adrenergic, and the central muscarinic receptor [Leslie et al., 1980]. Table 2 shows that also the specific 5-HT binding to a crude membrane fraction, 2 x 45,000 g ppt. [Enna et al., 1977], obtained from postmortem forebrain is decreased in the presence of ascorbate. The MAO inhibitor pargyline increases the specific 5-HT binding, perhaps by blockade of 5-HT catabolism. But other mechanisms may contribute to the increase in 5-HT binding, as will be discussed later. As compared with specific 5-HT binding obtained in "pure" 1 mM Tris-HCl buffer an almost identical binding is found if both compounds, ascorbate and pargyline, are present, though the 5-HT receptor system is likely affected in different ways by the various buffers.

Activation of 5-HT Binding
In a comparative study of microsomal, P_3 [Snyder and Bennett, 1975] and mitochondrial, P_2 [Wesemann, 1969] fractions of synaptic membranes it was ob-

served that tryptamine and 5-methoxytryptamine increase 5-HT binding specifically to the P_2 fraction
(Figure 1) while in accordance with Bennett and
Snyder and Bennett (1975) a displacement of 5-HT from the P_3
fraction is found. Since 5-HT binding is also

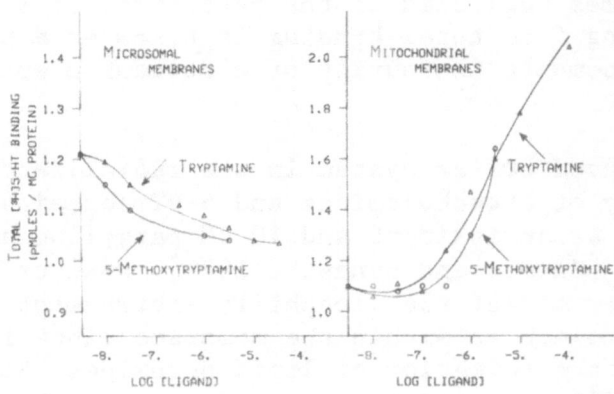

Figure 1 Effect of tryptamine and 5-methoxytrypta-
mine on [^{3}H]5-HT binding to microsomal and mitochondrial membranes from rat brain. - Membranes were incubated for 10 min at 37°C with 7 nM [^{3}H]5-HT with
tryptamine and 5-methoxytryptamine, respectively,
prior to the binding assay with the filtration procedure according to Snyder and Bennett (1975).

increased if crude membrane fractions, 2 x 45,000 g
ppt., isolated from rat brain and substances like
tryptamine, 5-methoxytryptamine, tyramine, ß-phenylethylamine, and ß-(4-methoxyphenyl)ethylamine are
used as sensitizing agents, a screening test was
performed with crude membranes obtained from post-
mortem forebrain. The results given in Table 3 are
preliminary since rather high concentrations of the
activating substances and only the forebrain of a
patient who died from bronchogenic carcinoma was
used. However, the increase in 5-HT binding parallels the effect observed with rat brain. Figure 2

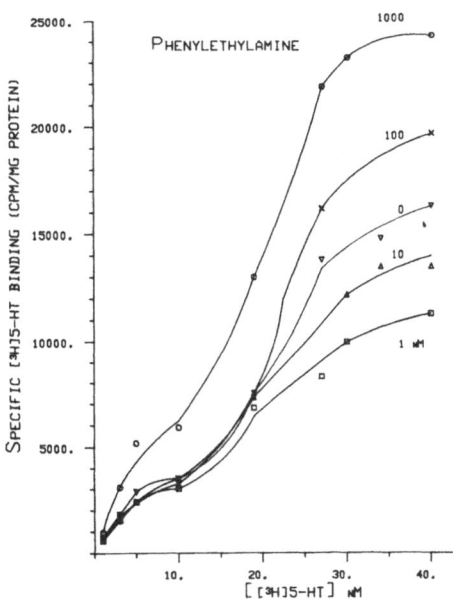

Figure 2   Dose-response curves of specific [^{3}H]5-HT
binding to postmortem forebrain (crude membrane
fraction) in the presence of various concentrations
of phenylethylamine. - Membranes were incubated for
10 min at 37°C concomitantly with [^{3}H]5-HT (1-40 nM)
and phenylethylamine (1-1000 nM).

shows the concentration-dependence of 5-HT binding
in the presence of phenylethylamine. At a concentration of 0.1 µM or higher phenylethylamine increases,
at a concentration of 10 nM or lower it decreases
5-HT binding. Thus, 5-HT binding to crude membranes
isolated either from human or from rat brain is
affected by substances which are possibly involved
in such brain dysfunctions as migraine, depression,
Huntington's chorea or hallucination.

Effect of MAO Inhibitors on the Activated 5-HT Binding

Since the activators of 5-HT binding are substrates of MAO-B, which catalyzes the oxidation of the amines

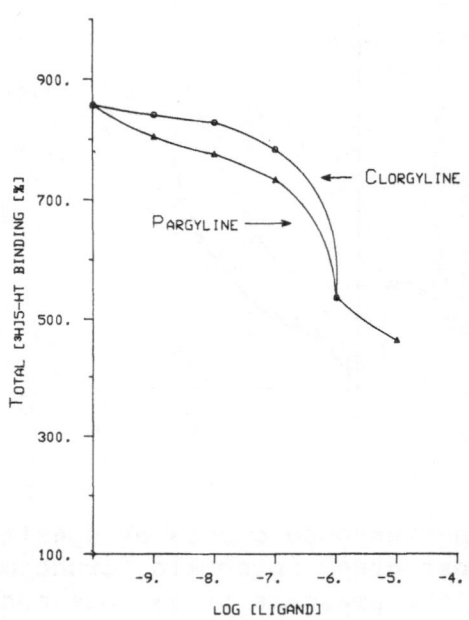

Figure 3 Decrease of tryptamine-induced activation of [³H]5-HT binding in rat brain by various concentrations of clorgyline and pargyline. - Crude membranes [Enna et al., 1977] were resuspended at 4°C in 1 mM Tris-HCl, pH 7.2, containing 0.1 mM tryptamine prior to the simultaneous addition of 30 nM [³H]5-HT and the MAO inhibitors; protein concentration: 550 μg/ml. Incubation and binding assay were performed as described in the legend of Table 1.

via the corresponding aldehydes to the acetic derivatives, phenylacetaldehyde, phenylacetic acid, 4-hydroxyphenylacetic acid, 3.4-dimethoxyphenyl-

TABLE 3. Activation of Specific $[^3H]$5-HT Binding to Human Forebrain and the Effect of Pargyline

Activator		Tris-HCl	Activation/Inhibition [%] Tris-HCl + 10 µM Pargyline
-		100	100
Tryptamine	1 µM	208	146
	10 µM	405	160
	100 µM	476	149
5-Methoxytryptamine	3.2 µM	337	146
	10.0 µM	425	134
	320.0 µM	567	136
Tyramine	1 µM	204	100
	10 µM	490	100
	100 µM	749	77

Crude membranes isolated from human forebrain [Enna et al., 1977] were resuspended in 1 mM Tris-HCl buffer, pH 7.2, and, in a parallel assay, in the same buffer containing 10 µM pargyline (protein conc.: 1500 µg/ml). Incubation was performed for 10 min at 37°C with 8.8 nM $[^3H]$5-HT in the presence and absence of the indicated activator. Each experiment was performed in quadruplicate.

63

acetic acid, D,L-4-hydroxymandelic acid, indoleacet-
aldehyde, indoleacetic acid, 5-HIAA, and 5-methoxy-
4-hydroxyphenylacetic acid are included in the
studies. None of the metabolites is as active as
sensitizing agent as the corresponding parent com-
pound; on the contrary, most of them are either inactive or displace 5-HT binding when applied at 0.1
mM concentration. Though tryptamine may act as a MAO
inhibitor it is unlikely that it sensitizes 5-HT
binding by MAO inhibition: (1) The effect of tryptamine exceeds by far that observed in the presence of
pargyline. (2) The effect of the MAO inhibitors
pargyline and clorgyline is not additive to tryptamine induced activation. On the contrary, they displace 5-HT after preceding increase of receptor
sensitivity by tryptamine (Figure 3). (3) Besides
other MAO inhibitors like deprenyl and iproniazid
the tuberculostatic agent isoniazid decreases tryptamine-induced activation of 5-HT binding though it
has no inhibitory action on MAO. In this context it
is interesting to mention that isoniazid and iproniazid, which have also been used in the treatment
of tuberculosis, induce in some patients psychotic
disorders. This side effect can not be explained by
MAO inhibition. Our present finding that MAO inhibitors and isoniazid reduce tryptamine sensitized
5-HT binding, suggests a new experimental approach
to this problem.

We have not proved by in vivo experiments the physiological relevance of the 5-HT sensitizing and de-
sensitizing system. However, besides the observation that Huntington's chorea and migraine can be
exacerbated by cheese, chocolate or red wine, which
contain tyramine or phenylethylamine, a few data
should be mentioned which suggest the involvement
of trace amines in neurotransmission: The excretion
of tryptamine and indoleacetic acid is enhanced when
the condition of patients with schizophrenia deter-
iorates [Brune and Himwich, 1963; Berlett et al.,
1964]; p-tyramine, m-tyramine, and β-phenylethylamine
potentiate the response of neurons in rat brain to

the inhibitory action of dopamine and noradrenaline
[Jones and Boulion, 1980]; subthreshold concentrations of tryptamine enhance the cholinergic transmission at the neuromuscular junction in frog
[Friedman, 1979]. The metabolism and/or the effect of
the amines is changed by treatment with MAO inhibitors or isoniazid: Nialamide increases tryptamine
excretion 3-5 fold [Friend et al.,1965]; isoniazid
ameliorates in some cases the symptoms of Huntington's chorea [Perry et al., 1979].

The problem is unsolved as to whether there exists an
endogenous analogue which replaces MAO inhibitors or
isoniazid modulating the sensitized 5-HT binding. In
a few pilot experiments a modulation of tryptamine-
induced increase of 5-HT receptor sensitivity was
observed using substances like morphine, ethylmor-
phine, nalorphine, naloxone, levallorphan, leu-
enkephalin, and met-enkephalin. Thus, attempts of
our laboratory appear to be justified with the aim
of elucidating the biological significance of these
results.

Effect of Sleep Deprival
A reduction of 5-HT binding to crude membranes,
2 x 45,000 g ppt., isolated from rat brain was observed if rats were deprived of sleep for 12 or 18
hours (group A: awake). Though 5-HT binding was decreased by $\pm$ 50% (Figure 4), the binding capacity returned towards control values (group C: control) if
rats were allowed to sleep for 1 hour prior to decapitation and binding measurements (group B: awake
+ 1 h sleep). After deprival for 24 hours no significant differences in 5-HT binding could be obtained
between group A, B and C. Though no explanation can
be given yet as to the nature of the stressor reducing 5-HT binding, an effect of different light/
dark periods can be excluded. In the experiment
with 12 h sleep deprival group A (awake) and group C
(control) were exposed to light from 7 a.m. to 7 p.m.
Only group B (awake + 1 h sleep) was in the dark
during the sleeping time.

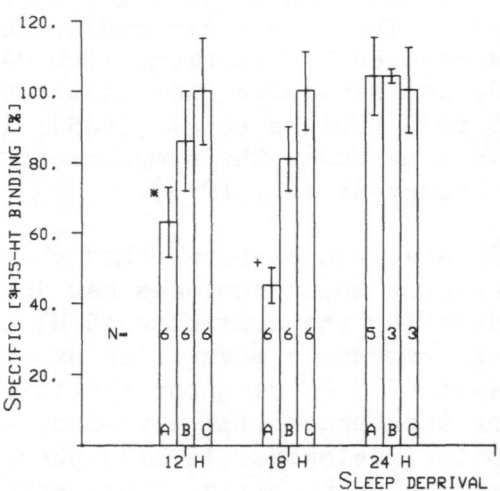

Figure 4   Specific [^{3}H]5-HT binding in rat brain
after sleep deprival. - Specific [^{3}H]5-HT binding to
crude membrane fractions (Enna et al., 1977) was
measured in 3 experimental groups: (A) awake for 12,
18 or 24 h prior to decapitation; (B) was allowed to
sleep for 1 h after sleep deprival; (C) control.
Incubation medium: 20 nM [^{3}H]5-HT in 1 mM Tris-HCl,
pH 7.2; protein conc.: 600-800 µg/ml.
^{x}p < 0.05 different from control group; $^+$p < 0.05
different from control group and awake + 1 h sleep
group.

CONCLUSION

As far as it could be tested, exogenous and endogenous factors affect 5-HT binding to central receptors in the same way in the human and in the rat.
In vitro experiments with crude membrane fractions
on the modulation of 5-HT receptor sensitivity
suggest the use of this system as a model to elucidate functional and pathological changes in neuro-

transmission explained so far as super-and subsensitivity of receptors. It might be possible that disturbances in receptor regulation is either the cause or the consequence of brain dysfunctions, drug treatment, or stress. Exogenous factors may influence as well the in vitro regulation as the radioligand-binding assay.

REFERENCES

Bernheimer, H., and Hornykiewicz, O. (1973). Brain amines in Huntington's chorea. In Advances in Neurology, Vol. 1 (eds. A. Barbeau, T.N. Chase, and G.W. Paulson), pp. 525-531, Raven Press, New York.

Berlett, H.H., Bull, C., Himwich, H.E., Kohl, H., Matsumoto, K., Pscheidt, G.R., Spaide, J., Tourlentes, T.T., and Valverde, J.M. (1964). Endogenous metabolic factor in schizophrenic behavior. Sci., 144, 311-313.

Birkmayer, W. (1969). Der α-Methyl-p-tyrosin-Effekt bei extrapyramidalen Erkrankungen. Wien. Klin. Wochenschr., 81, 10-12.

Brune, G.G., and Himwich, H.E. (1963). Biogenic amines and behavior in schizophrenic patients. Rec. Adv. Biol. Psychiat., 5, 144-160.

Del Castillo, J. and Katz, B. (1955). On the localization of acetylcholine receptors. J. Physiol. (Lond.), 128, 157-181.

Enna, S.J., Bird, E.D., Bennett, J.P.,Jr., Bylund, D.B., Yamamura, H.I., Iversen, L.L., and Snyder, S.H. (1976). Huntington's chorea. Changes in neurotransmitter receptors in the brain. New England J. Med., 294, 1305-1309.

Enna, S.J., Bennett, J.P.,Jr., Bylund, D.B.,Creese, I., Burt, D.R., Charness, M.E., Yamamura, H.I., Simantov, R., and Snyder, S.H. (1977). Neurotransmitter receptor binding: regional distribution in human brain. J. Neurochem., 28, 233-236.

Friedhoff, A.J., and Alpert, M. (1978). Receptor sensitivity modification as a potential treatment. In Psychopharmacology: a generation of progress (eds. M.A. Lipton, A. DiMascio, and K.F. Killam), pp. 797-801, Raven Press, New York.

Friedman, R.N. (1979). Tryptamine induced alterations of acetylcholine release at a neuromuscular juntion. Soc. Neurosci. Abstr. 5, 480.

Friend, D.G., Bell, W.R., and Kline, N.S. (1965). The action of L-dihydroxyphenylalanine in patients receiving nialamide. Clin. Pharmacol. Ther., 6, 362-366.

Jones, R.s.g. and Boulion, A.A. (1980). Interactions between p-tyramine, m-tyramine, or ß-phenylethylamine and dopamine on single neurons in the cortex and caudate nucleus of the rat. Can. J. Physiol. Pharmacol., 58, 222-227.

Leslie, F.M., Dunlap, C.E.,III, and Cox, B.M. (1980). Ascorbate decreases ligand binding to neurotransmitter receptors. J. Neurochem., 34, 219-221.

Perry, Th.L., Wright, J.M., Hansen, S., and MacLeod, P.M. (1979). Isoniazid therapy for Huntington's disease. In Advances in Neurology, Vol. 23 (eds. T.N. Chase, N.S. Wexler, and A. Barbeau), pp. 785-796, Raven Press, New York.

Savage, D.D., Mendels, J., and Frazer, A. (1980). Monoamine oxidase inhibitors and serotonin uptake inhibitors: differentiated effects on [^{3}H]serotonin binding sites in rat brain. J. Pharmacol. Exp. Ther., 212, 259-263.

Schaefer, A., Komlos, M., and Seregi, A. (1975). Lipid peroxidation as the cause of the ascorbic acid induced decrease of adenosine triphosphatase activities of rat brain microsomes and its inhibition by biogenic amines and psychotropic drugs. Biochem. Pharmacol., 24, 1781-1786.

Snyder, S.H. and Bennett, J.P.,Jr., (1975). Biochemical identification of the postsynaptic serotonin receptor in mammalian brain. In Modern pharmacology - toxicology, Vol. 3: Pre- and post-synaptic receptors (eds.E.Usdin and W.E.Bunney,Jr) pp. 191-206, Dekker, New York.

Stahl, W.L. and Swanson, P.D. (1974). Biochemical
abnormalities in Huntington's chorea brains.
Neurology, 24, 813-819.

Weiner, N., Martin, C., Wesemann, W., and Riederer,
P. (1979). Effect of post-mortem storage on
synaptic 5-HT binding and uptake in mammalian
brain. J.Neural Transm., 46, 253-262.

Wesemann, W. (1969). Isolation of 5-hydroxytrypta-
mine containing vesicles and of synaptic membranes from rat brain. FEBS Letters, 3, 80-84.

Section III
Enzymes and Receptors

Dopamine Receptors in Human Brain: Comparison with Rat and Calf Brain

STEPHEN LIST AND PHILIP SEEMAN

Department of Pharmacology,
University of Toronto,
Toronto, Canada M5S 1A8

In 1975, two dopaminergic binding sites were identified in the calf caudate nucleus, using the direct receptor binding assay method (Seeman et al., 1975; Burt et al., 1975). One site, which we now call the D2 dopamine receptor, has been labelled by ^{3}H-haloperidol and other ^{3}H-neuroleptics, as well as by the dopamine ergot agonists ^{3}H-dihydroergocryptine (^{3}H-DHEC), ^{3}H-lisuride and ^{3}H-bromocryptine (Leysen et al., 1978; Titeler et al., 1978; Martres et al., 1978; Fujita et al., 1979; Closse et al., in press). The D2 site is characterized by a high affinity (with 1-10 nM K_D) for a number of biologically potent neuroleptics, a high affinity for dopaminergic ergot agonists, such as bromocryptine and dihydroergocrytine (DHEC), and a lower affinity (with K_D values of 100-1000 nM) for dopaminergic catecholamines such as apomorphine and dopamine itself.

The second dopaminergic binding site, identified in the calf caudate, which we now refer to as the D3 site, can be labelled by ^{3}H-dopamine, ^{3}H-apomorphine or ^{3}H-ADTN. This site is characterized by its high affinity (with 1-10 nM KD) for dopamine as well as a number of other dopaminergic catecholamines, such as apomorphine and ADTN, an intermediate affinity for dopaminergic ergots and a low affinity (200-5000 nM K_D) for a number of potent neuroleptics (Burt et al., 1976; Seeman et al., 1976a).

It is now generally accepted that the D2 receptor is the site of action of classical postsynaptic dopamine receptor agonists and antagonists which elicit or block dopaminergic stereotypy and rotation, prolactin-secretion, and psychosis (Creese et al., 1976; Seeman et al., 1976b; Caron et al., 1978; Titeler and Seeman, 1978). The function of the D3 site is not yet known; however, it has been proposed that this site may function as a dopamine 'autoreceptor' (Titeler et al., 1978).

73

While the original characterization of the D_2 and D_3 sites were done in the bovine caudate, it is important to establish the existence of these dopaminergic sites in other species as well. Characterization of dopamine receptors in post-mortem human brain tissue is particularly important in   order to interpret properly changes in dopamine receptor binding occurring in human tissues under various treatment conditions or in pathological states. We have, therefore, been characterizing dopamine receptor binding in the human brain, as well as in calf and rat brain, in order to determine whether similar dopamine receptors are detectable in all three species.

D_2 Dopamine Receptors Labelled by ^{3}H-Spiperone in Human, Rat and Calf Brains:

In our laboratory we have tested a variety of neuroleptics on their ability to compete with the binding of 0.25 nM ^{3}H-spiperone in calf and human caudate, as well as rat striatum. Specific binding was defined as that binding inhibited by 1 μM (+)-butaclamol. The drug IC_{50} value was defined as the concentration of drug required to inhibit 50% of the specific ^{3}H-spiperone binding. As shown in Figure 1 below (which compares IC_{50} values of neuroleptics in the human caudate to the IC_{50} values of these drugs in the rat and calf) the human caudate shows a ^{3}H-spiperone binding site which is similar to that obtained in the calf and rat. The neuroleptic IC_{50} profile, shown in the figure, corresponds to the characteristics of the D_2 dopamine receptor.

Our present IC_{50} values for ^{3}H-spiperone binding in the calf and human caudate and rat striatum are consistent with those previously obtained in the calf and rat using ^{3}H-haloperidol to label the D_2 site. As has been noted previously, when using ^{3}H-spiperone the absolute drug IC_{50} values are somewhat higher than those obtained using ^{3}H-haloperidol (Leysen et al., 1978).

As can be seen in Figure 1, only one drug, (±)-sulpiride, lies significantly off the correlation curve. While (±)-sulpiride has a similar affinity for the D_2 site in the rat and human (with IC_{50} values against ^{3}H-spiperone of 200 nM and 400 nM, respectively), the drug has a 10-fold lower affinity (with an IC_{50} of 3000 nM) in the calf caudate. A 10-fold lower affinity for sulpiride on calf caudate D_2 receptors relative to rat and human has also been noted by Creese et al. (1979). While it has been shown that in general the affinities of neuroleptic drugs for the D_2 site in the calf correlate well with their clinical potencies in treating schizophrenia (Seeman et al., 1976b; Creese et al., 1976), the occasional drug such as sulpiride is more potent clinically than would be predicted by receptor binding in the calf (Lin et al., 1980). Sulpiride's affinity for the D_2 site in the human, however, is much more compatible with its observed clinical potency.   Therefore, while the D_2 receptors of rat, calf and hu-

74

man are generally similar, there appear to be some differences and
it is best to compare biological drug potencies and drug receptor
affinities within the same species.

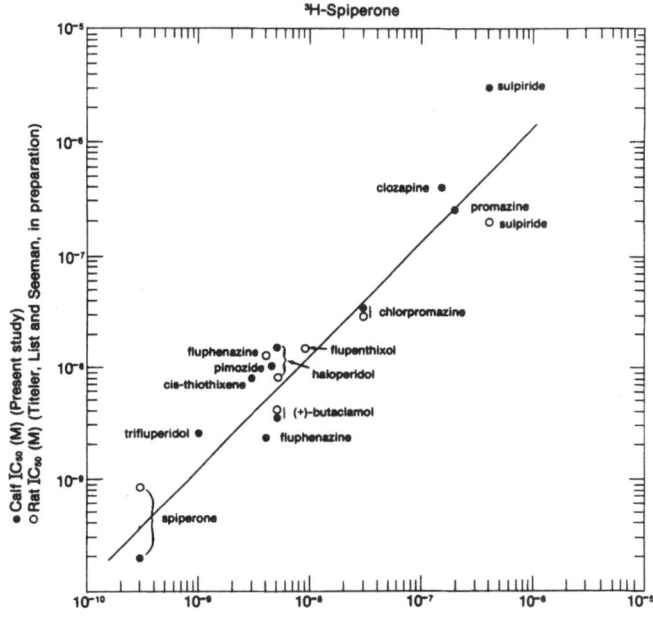

Fig. 1. The IC50 values (in moles/liter) for various drugs
against the specific binding of 3H-spiperone in the human caudate
(Lee and Seeman, 1979), (horizontal axis) agree and correlate
with those obtained in the rat striatum (Titeler, List and
Seeman, manuscript in preparation) and the calf caudate (present
study). The correlation coefficients of drug IC50 values for
human vs. calf and human vs. rat were 0.93 and 0.96, respectively.

D3 Dopamine Receptors Labelled by 3H-Dopamine in the Human, Calf and Rat Brain

As has been discussed in the previous section, there is a
general consistency between laboratories and species on the
characteristics of the binding site labelled by 3H-spiperone in
the caudate (i.e. the D_2 dopamine receptor).

While many laboratories have labelled the other dopaminergic
site (the D_3 site) in the calf brain using 3H-dopamine, 3H-apomor-
phine, and 3H-ADTN (Seeman et al., 1976a, 1978, 1979; Burt et
al., 1976; Cross et al., 1979) a number of studies have failed
to label this site in the rat brain (Burt et al., 1976; Creese
et al., 1978; Leysen, 1980). For example, while Burt et al.
(1975) labelled a site in the rat using 3H-dopamine, this site

had only a low affinity for dopamine, with an IC_{50} of 380 nM, which is considerably lower than the D_3 site IC_{50} for dopamine of 3-10 nM. Creese et al. (1978) using ^{3}H-apomorphine and ^{3}H-ADTN labelled sites in the rat which had high affinity for the radioligands, but again, which had only low affinity for dopamine. As shown in Figure 2, drug IC_{50}'s against ^{3}H-dopamine or ^{3}H-apomorphine binding in the rat, in the studies of Burt et al. (1975) and Creese et al. (1978) show no correlation with the typical drug IC_{50} profile of the D_3 site labelled by these ligands in the calf caudate. Furthermore, it was not reported in these studies

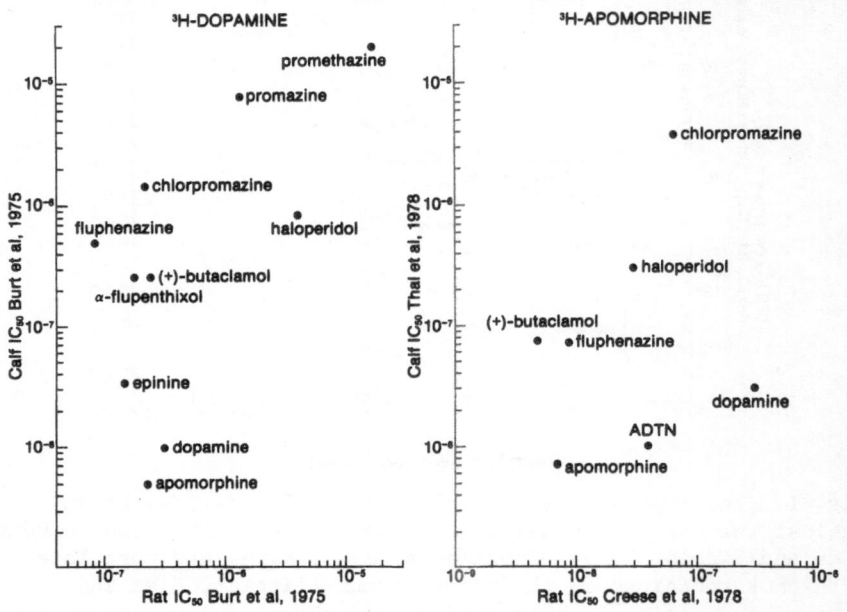

Fig. 2. The data on the left indicate that in the study of Burt et al., 1975 there is a marked difference between the IC_{50}'s of various drugs against ^{3}H-dopamine binding in the rat striatum (horizontal axis) and those obtained in the cow caudate. Only the cow ^{3}H-dopamine binding site shows D_3 type binding. The data on the right indicate that there is no agreement between the ^{3}H-apomorphine IC_{50} values observed in cow caudate by Thal, Creese and Snyder (1978) and those observed in the rat by Creese, Prosser and Snyder (1978). Only the cow ^{3}H-apomorphine IC_{50} values are compatible with D_3 type dopamine binding.

whether the sites labelled in the rat had a distribution corresponding to dopaminergic areas of the brain. In fact, there is no indication in these studies that the sites in the rat that were labelled by ^{3}H-dopamine and ^{3}H-apomorphine were even dopaminergic in nature.

Some other laboratories, however, appear to have labelled the D_3 dopamine site in the rat using 3H-apomorphine and ^{3}H-dopamine, although the site was in most cases not extensively characterized (Goldstein et al., 1978; Blackburn et al., 1978). Furthermore, Creese et al. (1979) while unable to detect the D_3 site in the rat, labelled a site with D_3 properties in the human brain, using 3H-apomorphine.

Many discrepancies between laboratories and species may be accounted for on the basis of differences in binding assay conditions (Titeler et al., 1979). As our laboratory (and other laboratories using assay conditions similar to ours) are presently involved in studying 3H-dopamine and ^{3}H-apomorphine binding in pathological post-mortem human brains and in rats which have undergone various drug treatments, we have been fully characterizing ^{3}H-dopamine and 3H-apomorphine binding in the human and the rat in order to determine whether the same D_3 dopamine site present in the calf is also labelled in these other species.

Using ^{3}H-dopamine concentrations of 0.1 nM – 5 nM (and 1 μM cold apomorphine to define specific binding), Scatchard analyses were performed in a variety of regions of the human, calf and rat brain.

Analysis of five separate human brains revealed a high affinity ^{3}H-dopamine binding site in the caudate and putamen with average K_D values of 2.2 ± 0.4 nM, and 2.8 nM ± 0.3 nM, respectively. These K_D's are similar to those observed in calf caudate and rat striatum of 2.5 nM and 2.3 nM respectively (Titeler et al., 1979). The average densities of the ^{3}H-dopamine sites in the human caudate and putamen were 63 ± 10 fmoles/mg protein and 57 ± 11 fmoles/mg protein, respectively (List et al., 1980). These densities are similar to the number of sites in the rat striatum (81 fmoles/mg) and about 2.5-fold smaller than the number of high-affinity 3H-dopamine binding sites observed in the calf caudate (180-200 fmoles/mg). Figure 3 shows a representative Scatchard analysis of one human caudate.

From preliminary work it appears that the ^{3}H-dopamine binding site in the human brain has a dopaminergic distribution with high-affinity sites detectable in the caudate and putamen (dopamine-rich areas) while dopamine-poor areas (globus pallidus, thalmus and hippocampus) showed no detectable high affinity 3H-dopamine binding. We have also found a dopaminergic distribution for rat 3H-dopamine binding as well, with high-affinity sites in the dopamine-rich areas, striatum and olfactory tubercle, while in dopamine-poor areas (cerebellum, thalamus, hippocampus, hypothalamus and frontal cortex) no high-affinity binding was detectable.

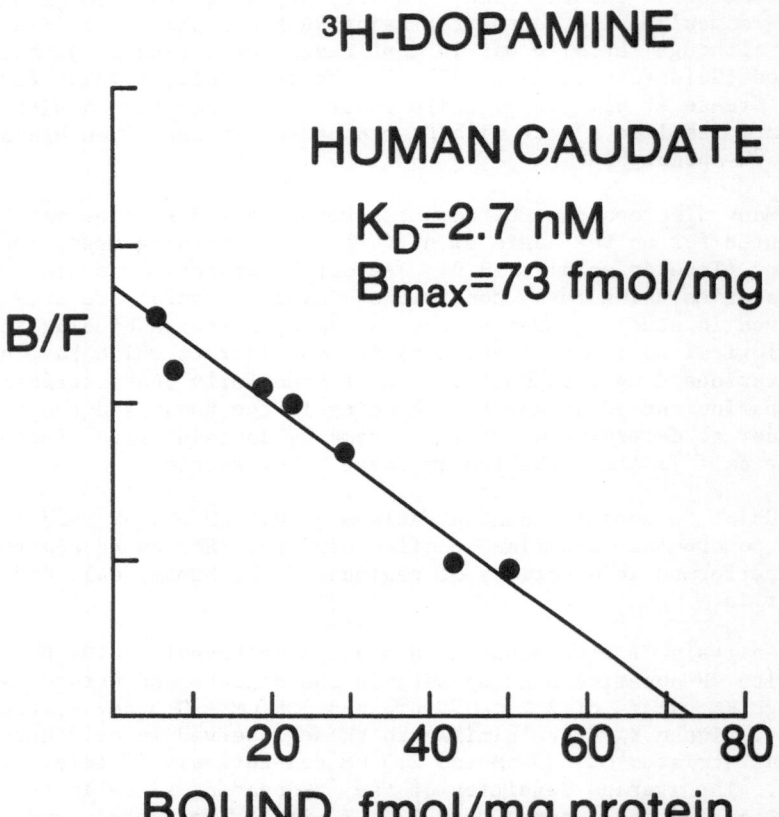

^{3}H-DOPAMINE

HUMAN CAUDATE
K_D=2.7 nM
B_{max}=73 fmol/mg

B/F

BOUND, fmol/mg protein

Fig. 3. Scatchard analysis of specific ^{3}H-dopamine binding (defined by 1 µM apomorphine) in a single representative human caudate.

We have tested a variety of drugs for their ability to compete with the specific binding of 0.7 nM ^{3}H-dopamine in the
human and calf caudate and the rat striatum.

Fig. 4 shows competition curves of various drugs against
^{3}H-dopamine binding in an individual human caudate, and Table 1
(third column) lists average ^{3}H-dopamine drug IC_{50} values obtained from determinations on a number of (5-10) individual
human caudates. As demonstrated by Figure 4 and Table 1, the
^{3}H-dopamine binding in the human caudate appears to be to the
D_3 dopamine site, as the dopaminergic catecholamines (dopamine,
apomorphine, ADTN and N-propyl-norapomorphine) had IC_{50} values
in the high-affinity range (3-6 nM), while the neuroleptics had
IC_{50} values in the low affinity range (700-7000 nM) and the dopaminergic ergots had intermediate IC_{50} values in between the two
(in the 30-500 nM region).

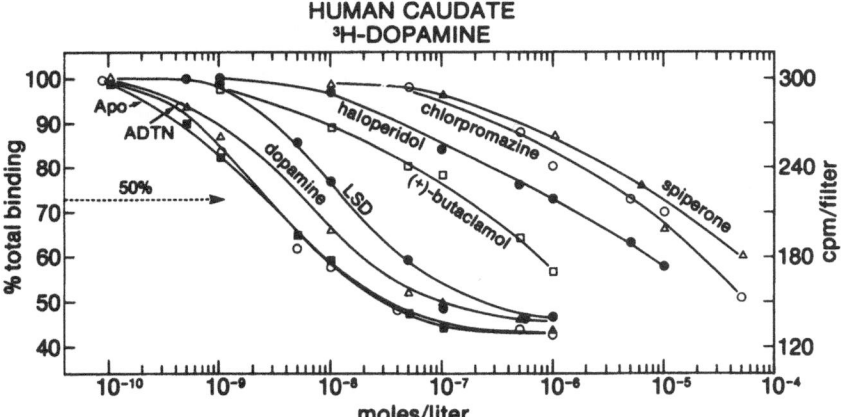

Fig. 4. Competition of dopaminergic catecholamines, ergots and neuroleptics against the binding of 0.7 nM ^{3}H-dopamine to an individual human caudate. Specific ^{3}H-dopamine binding was defined as that inhibited by 1 μM apomorphine and was 55% of the total binding. (Apo = apomorphine; ADTN = 6,7-dihydroxy-2-aminotetralin).

Table 1

[3]H-Dopamine IC50 Values (nM)

Drug	Rat Striatum (Ref. 27)	Calf Caudate (This study)	Human Caudate (Ref. 16)
±-ADTN	1.5	1.5	5.1
Apomorphine	3.5	4	3.4
Dopamine	5	3	4.3
(±)-NPA	6	4.5	3.9
d-LSD	13	12	17
(-)-Norepinephrine	–	20	28
DHEC	100	75	60
Bromocryptine	400	300	437
(+)-Butaclamol	160	160	720
Haloperidol	200	800	3000
Chlorpromazine	500	2500	6500
Fluphenazine	500	–	–
Spiperone	1200	2000	5100

All experiments used 0.7 nM ^{3}H-dopamine and 1 μM apomorphine to define specific binding. (±)-ADTN = (±)-6,7-dihydroxy-2-amino-tetralin, DHEC = dihydroergocryptine, (±)-NPA = (±)-N-propyl-norapomorphine.

As shown in Table 1, drug IC50 values against ^{3}H-dopamine binding in rat striatum show an order of potency similar to those in the human, indicating again D3 type dopamine binding.

79

As shown in Figure 5, the pharmacological profile of the D3 site in the human caudate correlates well with the D3 site in the rat and calf labelled by ^{3}H-dopamine. As can be seen in Table 1 some differences in the D3 receptors of the three species do exist, however. While catecholamine dopamine agonists and dopaminergic

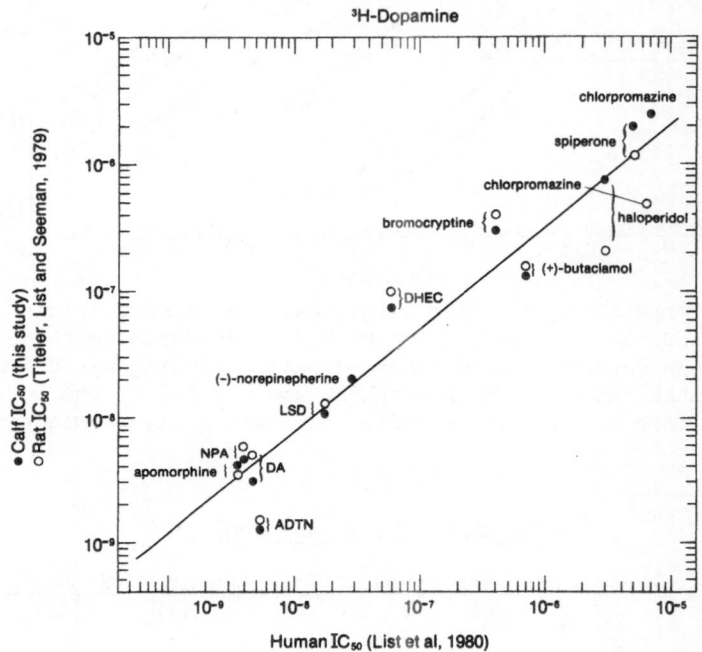

Fig. 5. The IC_{50} values for various drugs against the specific binding of ^{3}H-dopamine to human caudate (List et al., 1980) (horizontal axis) correlate with the values obtained in the rat striatum (Titeler, List and Seeman, 1979) and the calf caudate (present study) (vertical axis). The correlation coefficients of drug IC_{50} values for human vs. calf and human vs. rat were 0.98 and 0.94, respectively.

ergots show similar affinities for the D3 site across all species, relative to the calf caudate, neuroleptics tend to have a higher affinity for the rat D3 site and a somewhat lower affinity for the D3 site in man.

From our studies it is clear that, in our laboratory, using ^{3}H-dopamine, we can label a dopamine receptor (the D3 site) in the human brain similar to that of the calf and the rat. As we have shown, this dopamine binding site has a high affinity for dopaminergic agonists, has a dopaminergic distribution, and is present in three different species. We recommend that this site be termed the D3 dopamine receptor. As has been discussed, the same consistency in binding between species has not been obtained in all laboratories. It appears that under certain assay conditions the D_3 site can be labelled in some species while not in others (where the radioligand may bind to completely different and rather ill-defined sites).

In order to interpret changes in dopamine receptor binding occurring in pathological brain tissue or as the result of drug treatment in man and animal it is important to know what site the radioligand is binding to. We have observed, for example, that these D_3 sites were about 50% reduced in the striatum of Parkinson's diseased brains (Lee et al., Nature, 1978), as measured by ^{3}H-apomorphine, or as measured by ^{3}H-dopamine (Seeman and Lee, 1980), a finding compatible with the idea that at least 50% of the D_3 sites are presynaptic. Finally, in order to compare results between laboratories on either 3H-apomorphine or 3H-dopamine, it is important to establish that the radioligand is labelling the same site in both laboratories, since the experimental conditions often differ.

References

1. Blackburn, K.J., Bremner, R.M., Greengrass, P.M. and Morville, M. (1978). Selective affinities of bromocriptine and lergotrile for rat limbic dopamine binding sites. Proc. Brit. Pharmacol. Soc. 413P.

2. Burt, D.R., Enna, S.J., Creese, I. and Snyder, S.H. (1975). Dopamine receptor binding in the corpus striatum of mammalian brain. Proc. Nat. Acad. Sci. 72, 4655-4659.

3. Burt, D.R., Creese, I. and Snyder, S.H. (1976). Properties of ^{3}H-haloperidol and ^{3}H-dopamine binding associated with dopamine receptors in calf brain membranes. Mol. Pharmacol. 12, 800-812.

4. Caron, M.G., Beaulieu, M., Raymond, V., Gagne, B., Drouin, J., Lefkowitz, R.J. and Labrie, F. (1978). Dopaminergic receptors in the anterior pituitary gland: Correlation of 3H-DHEC binding with the dopaminergic control of prolactin release. J. Biol. Chem. 253, 2244-2253.

5. Closse, A., Frick, W., Hauser, D. and Sauter, A. Characterization of ^{3}H-bromocriptine to calf caudate membranes. In Psychopharmacology and Biochemistry of Neurotransmitters (eds. H. Yamamura, R. Olsen and E. Usdin), Elsevier North Holland, New York (in press).

6. Creese, I., Burt, D.R. and Snyder, S.H. (1976). Dopamine receptors and average clinical doses. Science 194, 545-546.

7. Creese, I., Prosser, T. and Snyder, S.H. (1978). Dopamine
    receptor binding: Specificity, localization and regulation
    by ions and guanyl nucleotides. Life Sci. $\underline{23}$, 495-500.

8. Creese, I., Stewart, K. and Snyder, S.H. (1979). Species
    variations in dopamine receptor binding. Eur. J. Pharmacol.
    $\underline{60}$, 55-66.

9. Cross, A.J., Crow, T.J. and Owen, F. (1979). The use of ADTN
    as a ligand for brain dopamine receptors. Brit. J. Pharmacol.
    $\underline{66}$, 87P-88P.

10. Fujita, N., Kihachi, S., Yonehara, N., Watanabe, Y. and
    Yoshida, H. (1979). Binding of ^{3}H-lisuride hydrogen maleate
    to striatal membranes of rat brain. Life Sci. $\underline{25}$, 969-974.

11. Goldstein, M., Lew, J.Y., Nakamura, S., Battista, A.F.,
    Lieberman, A. and Fuxe, K. (1978). Dopaminephilic properties
    of ergot alkaloids. Fed. Proc. $\underline{37}$, 2202-2206.

11A. Lee, T. and Seeman, P. (1979). Correlation of anti-
    psychotic drug potency and neuroleptic receptor inhi-
    bition in postmortem human brains. Soc. Neurosci.
    (Abstract) $\underline{5}$, 653

12. Lee, T., Seeman, P., Rajput, A., Farley, I.J. and Hornykiewicz,
    O. (1978). Receptor basis for dopaminergic supersensitivity
    in Parkinson's disease. Nature $\underline{273}$, 59-61.

13. Leysen, J.E., Gommeren, W. and Laduron, P.M. (1978). Spiperone:
    A ligand of choice for neuroleptic receptors. I. Kinetics and
    characteristics of in vitro binding. Biochem. Pharmacol. $\underline{27}$,
    307-316.

14. Leysen, J.E. (1980). Unitary dopaminergic receptor composed
    of cooperativity linked agonist and antagonist sub-unit
    binding sites. Comm. Psychopharmacol. $\underline{3}$, 397-410.

15. Lin, C.W., Maayani, S. and Wilk, S. (1980). The effect of
    typical and atypical neuroleptics on binding of ^{3}H-spiro-
    peridol in calf caudate. J. Pharmacol. Exper. Therap. $\underline{212}$,
    462-468.

16. List, S., Titeler, M. and Seeman, P. (1980). High-affinity
    ^{3}H-dopamine receptors (D_3 sites) in human and rat brain.
    Biochem. Pharmacol., in press.

17. Martres, M.P., Baudry, M. and Schwartz, J.C. (1978). Charac-
    terization of ^{3}H-domperidone binding on striatal dopamine
    receptors. Life Sci. $\underline{23}$, 1781-1784.

18. Seeman, P. and Lee, T. (1980). Brain dopamine receptors (D_2 and D_3 sites) in Parkinson's disease and schizophrenia. Proc. 12th Collegium. Int. Neuropsychopharmacol. (Göteborg), in press.

19. Seeman, P., Chau-Wong, M., Tedesco, J. and Wong, K. (1975). Brain receptors for antipsychotic drugs and dopamine: Direct binding assays. Proc. Nat. Acad. Sci. U.S.A. 72, 4376-4380.

20. Seeman, P., Lee, T., Chau-Wong, M., Tedesco, J. and Wong, K. (1976a). Dopamine receptors in human and calf brains, using ^{3}H-apomorphine and an antipsychotic drug. Proc. Nat. Acad. Sci. U.S.A. 73, 4354-4358.

21. Seeman, P., Lee, T., Chau-Wong, M. and Wong, K. (1976b). Antipsychotic drug doses and neuroleptic/dopamine receptors. Nature 261, 717-719.

22. Seeman, P., Tedesco, J.L., Lee, T., Chau-Wong, M., Muller, P., Bowles, J., Whitaker, P.M., McManus, C., Tittler, M., Weinreich, P., Friend, W.C. and Brown, G.M. (1978). Dopamine receptors in the central nervous system. Fed. Proc. 37, 130-136.

23. Seeman, P., Woodruff, G.N. and Poat, J.A. (1979). Similar binding of ^{3}H-ADTN and ^{3}H-apomorphine to calf brain dopamine receptors. Eur. J. Pharmacol. 55, 137-142.

23A.Thal, L., Creese, I. and Snyder, S. H. (1978). Tritiated apomorphine interactions with dopamine receptors in calf brain. Eur. J. Pharmacol. 49, 295-299

24. Titeler, M. and Seeman, P. (1978). Antiparkinsonian drug doses and neuroleptic receptors. Experientia 34, 1490-1492.

25. Titeler, M., Weinreich, P. and Seeman, P. (1977). New detection of brain dopamine receptors with ^{3}H-dihydroergocryptine Proc. Nat. Acad. Sci. U.S.A. 74, 3750-3753.

26. Titeler, M., Tedesco, J.L. and Seeman, P. (1978). Selective labelling of pre-synaptic receptors by 3H-dopamine, ^{3}H-apomorphine and ^{3}H-clonidine, labelling of post-synaptic sites by 3H-neuroleptics. Life Sci. 23, 587-592.

27. Titeler, M., List, S. and Seeman, P. (1979). High affinity dopamine receptors (D_3) in rat brain. Comm. Psychopharmacol. 3, 411-420.

Neurotransmitter Enzymes and Receptors in Post-mortem Brain in Schizophrenia: Evidence that an Increase in D_2 Dopamine Receptors is Associated with the Type I Syndrome

T. J. CROW, F. OWEN, A. J. CROSS, N. FERRIER, E. C. JOHNSTONE,
R. M. McCREADIE, D. G. C. OWENS and M. POULTER
Division of Psychiatry, Clinical Research Centre,
Northwick Park Hospital,
Harrow, HA1 3UJ,
England

The functional psychoses, including schizophrenia, were distinguished from the organic psychoses by Kraepelin and Bleuler, who considered that intellectual impairment was not a significant feature of the former group of illnesses. However, some recent work suggests that defects of temporal orientation, often considered a hallmark of the organic psychoses, do occur in schizophrenia (Crow and Mitchell, 1975; Crow and Stevens, 1978) and that in some chronic schizophrenic states intellectual impairments are associated with structural changes in the brain (Johnstone, Crow, Frith, Husband and Kreel, 1976). Such conditions may be an exception rather than the rule and it is commonly assumed that the defect in most schizophrenic illnesses has a chemical rather than an anatomical basis. It is plausible that the primary change is a disturbance of neurohumoural transmission, and in the past 25 years a number of hypotheses have been proposed (Table 1).

TABLE 1   Neurohumoural Hypotheses of Schizophrenia

Serotonin deficiency	Gaddum, 1954
Dopamine neurone overactivity	Randrup and Munkvad, 1967
Noradrenaline neurone degeneration	Stein and Wise, 1971
Monoamine oxidase deficiency	Murphy and Wyatt, 1972
GABA deficiency	Roberts, 1972
Dopamine receptor supersensitivity	Bowers, 1974

Each of these hypotheses has now been evaluated in a series of studies on brain tissue collected at post-mortem and evaluated with standard enzymatic, gas-chromatographic and ligand-binding techniques at Northwick Park. A series of over 60 brains of patients who were reported to have suffered from a psychotic illness have been collected, and among the most recent brains have been a series of 15 patients who had been assessed by ourselves in life with standard clinical rating schedules.

Monoamine Enzymes

In an early study we evaluated the hypothesis that noradrenaline neurones degenerate by assessing dopamine-β-hydroxylase (DBH) activity in 12 schizophrenic and 12 control brains (Table 2).

TABLE 2. DBH Activity in Post-mortem Brain of Schizophrenics and Controls (from Cross et al., 1978).

Brain Area	DBH Activity* (mean ± S.D.)			
	Controls (n=12)	Schizophrenics (n=12)	t	P
Hypothalamus	131 ± 41	156 ± 55	1.32	NS
Hippocampus	24 ± 8	25 ± 9	0.22	NS
Parietal cortex	25 ± 7	21 ± 8	1.36	NS
Frontal cortex	24 ± 7	27 ± 9	0.86	NS
Occipital cortex	27 ± 8	24 ± 7	0.96	NS
Temporal cortex	24 ± 6	25 ± 6	0.40	NS

* as nmol product formed/g tissue/h.

There were no significant differences between schizophrenics and controls. Catechol-O-methyl transferase activity was also found to be similar in these two sets of brains.

To investigate the hypothesis (Murphy and Wyatt, 1972) that monoamine oxidase activity is reduced we assessed the activity of this enzyme in fourteen brain areas using four substrates, serotonin as a substrate for the type A form of the enzyme, benzylamine for the type B form, tyramine and dopamine, which it has been suggested may be metabolised by a third form of the enzyme. The findings (Table 3) revealed no significant difference between schizophrenics and controls in any of the 56 comparisons.

Therefore, whatever may be the situation in the platelet these findings give no support to the view that a deficiency or abnormality of monoamine oxidase can account for the symptoms of schizophrenia.

TABLE 3   MAO Activities (nmol/mg protein/hr) in Post-mortem Brains from Controls and Schizophrenics (from Cross, Crow, Glover, Lofthouse, Owen and Riley, 1977).

mean ± s.d.

Brain Region	5HT		Benzylamine		Tyramine		Dopamine	
	C	S	C	S	C	S	C	S
Temporal cortex	36 ± 5	41 ± 10	43 ± 6	40 ± 5	100 ± 22	94 ± 19	15 ± 3	14 ± 3
Parietal cortex	37 ± 5	41 ± 6	30 ± 5	31 ± 8	97 ± 15	89 ± 12	18 ± 5	21 ± 6
Frontal cortex	44 ± 5	45 ± 4	41 ± 6	42 ± 6	121 ± 28	112 ± 14	28 ± 6	25 ± 4
Occipital cortex	56 ±16	49 ± 10	46 ±11	40 ± 9	98 ± 25	91 ± 14	30 ±10	26 ± 4
Cingulum	53 ±16	54 ± 16	68 ±19	67 ±16	193 ± 61	211 ± 25	19 ±10	20 ± 6
Hypothalamus	82 ±21	70 ± 13	131 ±32	117 ±19	214 ± 45	183 ± 36	34 ± 8	32 ±10
Caudate	47 ± 6	47 ± 5	86 ±15	92 ± 8	145 ± 27	145 ± 17	30 ± 4	28 ± 5
Putamen	44 ± 4	43 ± 6	85 ±12	75 ±12	137 ± 34	133 ± 22	17 ± 5	20 ± 7
Accumbens	68 ± 7	74 ± 7	130 ±24	145 ±31	189 ± 23	195 ± 35	51 ± 5	51 ±10
Thalamus	51 ± 4	48 ± 8	59 ±12	60 ±10	112 ± 15	113 ± 13	28 ± 5	27 ± 4
S. nigra	56 ± 7	59 ± 16	58 ±12	63 ±10	113 ± 18	133 ± 28	14 ± 4	15 ± 4
Cerebellum	25 ± 8	31 ± 7	16 ± 5	14 ± 6	38 ± 6	35 ± 5	4 ± 1	4 ± 1
Amygdala	71 ±11	72 ± 9	76 ±10	71 ±10	196 ± 23	183 ± 26	19 ± 3	21 ± 2

C = Controls (n=10)   S = Schizophrenics (n=9)

p is NS for any substrate.

GAD and GABA Binding

The hypothesis that glutamic acid decarboxylase (GAD) is deficient was also not supported (Table 4).

TABLE 4   GAD Activity (μmol/g/hr) in Post-mortem Brains of Controls and Schizophrenics

| | mean ± s.d. | | |
	Controls	Schizophrenics	'p'
Accumbens	4.9 ± 0.6 (18)	5.5 ± 1.0 (16)	NS
Caudate	3.6 ± 0.5 (18)	4.0 ± 0.6 (18)	NS
Putamen	3.3 ± 0.6 (18)	4.2 ± 0.7 (18)	NS
Amygdala	2.1 ± 0.3 (16)	2.5 ± 0.3 (18)	NS

Numbers in parentheses indicate brains tested.

GABA receptors were assessed by a ligand binding technique (using ^{3}H-GABA at a concentration of 10 nm) along with benzodiazepine receptors (assessed with ^{3}H-diazepam at a concentration of 1.6 nm) in caudate nucleus and putamen of 20 schizophrenic brains and 19 controls (Figure 1).

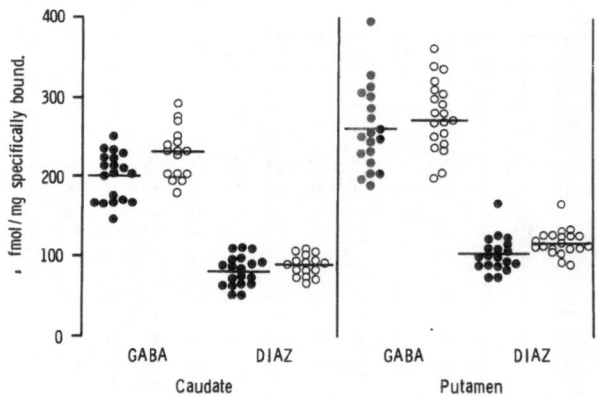

Figure 1.   GABA and Diazepam Binding in Post-mortem Brains

No systematic differences in schizophrenic by comparison with control brains were observed. Thus in these samples of patients there was no evidence to support the hypothesis of a disturbance of gabaminergic mechanisms, nor for the noradrenaline neurone degeneration or monoamine oxidase deficiency theories.

The Serotonin Hypothesis

Serotonergic transmission has been investigated with the use of ligands for the 5-HT receptor. ^{3}H-5HT and ^{3}H-LSD binding are significantly different from control brains (Figure 2)

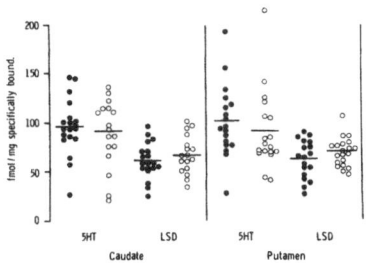

Figure 2. The Binding of ^{3}H-5HT and ^{3}H-LSD in Schizophrenic and control brain. (From Owen et al., 1980; Crow et al., 1980).

Tryptophan metabolism has also been found unchanged in schizophrenic brain (Joseph, Baker, Crow, Riley and Risby, 1979).

The Dopamine Hypothesis

By reason of the accumulating evidence on the mode of action of antipsychotic drugs by far the most compelling theory in recent years has been the dopamine hypothesis - that in some forms of schizophrenia there is either increased dopamine neurone activity (Randrup and Munkvad, 1966) or supersensitivity of dopamine receptors (Bowers, 1974; Crow, Deakin, Johnstone and Longden, 1976). In collaboration with workers in the MRC Neurochemical Pharmacology Unit in Cambridge we assessed the concentrations of dopamine (DA) and its metabolites homovanillic acid (HVA) and dihydroxyphenylacetic acid (DOPAC) in a total of 43 schizophrenics and 44 control brains (Table 5).

TABLE 5   Dopamine and its Metabolites in Two Series of Schizophrenic Brains by Comparison with Controls.   mean ± s.e.m.

Northwick Park Series	n = 19	n = 18
Caudate		
DA µg/g	1.6 ± 0.3	* 2.5 ± 0.3
HVA	5.4 ± 0.3	**3.8 ± 0.5
DOPAC	1.3 ± 0.18	0.8 ± 0.13
Accumbens		
DA	0.9 ± 0.3	0.7 ± 0.1
HVA	4.7 ± 0.5	5.5 ± 0.5
DOPAC	1.5 ± 0.02	1.1 ± 0.15
Cambridge Series	n = 25	n = 25
Caudate		
DA	1.7 ± 0.2	2.0 ± 0.2
HVA	4.3 ± 0.4	5.6 ± 0.8
DOPAC	0.8 ± 0.10	0.5 ± 0.10
Accumbens		
DA	1.4 ± 0.1	***2.0 ± 0.1
HVA	4.4 ± 0.3	4.9 ± 0.6
DOPAC	0.4 ± 0.04	0.4 ± 0.05

*p<0.05   **p<0.02   ***p<0.01

Although there was evidence for an increase in dopamine concentrations (in the nucleus accumbens in the Cambridge series, and caudate nucleus in the Northwick Park series) this was not a consistent finding. Moreover there was no evidence of overactivity of dopamine neurones as reflected by concentrations of the dopamine metabolites. Indeed the only significant difference was a decrease in HVA concentration in caudate nucleus in the schizophrenics in the Northwick Park series but this finding was not replicated in the Cambridge series. Overall the findings (Bird et al., 1979) give no support to the dopamine neurone overactivity hypothesis.

Dopamine Receptors

On the basis of earlier data (e.g. from c.s.f. studies) it had been predicted (Bowers, 1974; Crow, Deakin, Johnstone and Longden, 1976) that a disturbance in dopaminergic transmission might result from an increase in numbers of dopamine receptors rather than a change in presynaptic dopamine neurone activity. Increased butyrophenone binding in post-mortem schizophrenic brain was reported with spiroperidol (Owen, Cross, Crow, Longden, Poulter and Riley, 1978) and haloperidol (Lee, Seeman, Tourtellotte, Farley and Hornykiewicz, 1978). Although no change in spiroperidol binding was reported by some authors (e.g. Mackay, Doble, Bird, Spokes, Quik and Iverson, 1978) it appears that the quantities of neuroleptic remaining in the brain after death interfere with ligand binding (Owen, Cross, Poulter and Waddington, 1979). This effect may lead to spuriously low values in schizophrenic brain and only when relatively high ligand concentrations are used or maximum binding value is derived from a saturation analysis can the changes be demonstrated (Owen et al., 1978). Increased spiroperidol binding is present in caudate nucleus, putamen and nucleus accumbens (Figure 3).

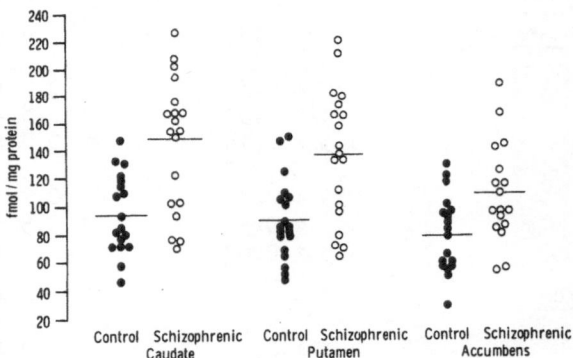

Figure 3. Dopamine receptor density (assessed with ^{3}H-spiroperidol as ligand at 0.8 nm concentration) in three dopaminergically innervated areas of brain. The differences between schizophrenics and controls are significant ($p < 0.001$) in caudate and putamen and ($p < 0.01$) in nucleus accumbens (from Owen et al., 1978).

Assessed as maximum binding in caudate nucleus (Figure 3) the values for two-thirds of patients fall outside the control range, and there is a mean increase of 103% (t = 5.15, p<0.001) in the schizophrenic brains.

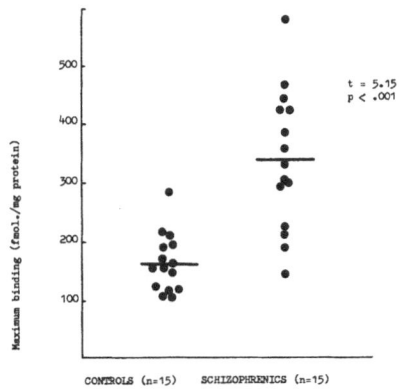

Figure 4. Dopamine receptor density (assessed as maximum specific spiroperidol binding) in caudate nucleus (from Owen et al., 1978).

It has been argued that the increase in binding in schizophrenic brain is not a secondary consequence of neuroleptic medication since a significant increase is found in some patients who have been free of such medication for the year before death (Owen et al., 1978). Recent studies of brain specimens from patients with Huntington's chorea have shown that patients who have been on neuroleptic medication in quantities apparently similar to those prescribed to some schizophrenic patients do not have increased spiroperidol binding by comparison with patients who have not had such medication. (Owen, Cross, Crow and Waddington, 1980).

It is also of interest that the binding of the dopamine agonist ADTN is not increased in schizophrenic brain (Figure 5).

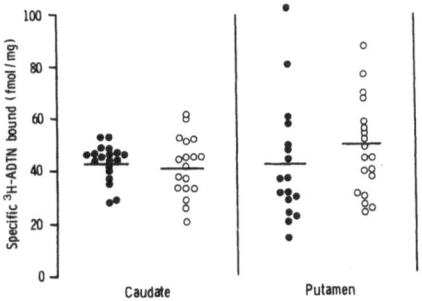

Figure 5. ADTN binding in post-mortem brain (from Owen et al., 1980 and Crow et al., 1980).

The binding of apomorphine, which is closely similar to that of ADTN, has also been found to be unchanged in schizophrenic brain (Lee et al., 1978) although the binding of this ligand is increased following chronic neuroleptic administration in animal experiments (Muller and Seeman, 1977). This suggests that the change in schizophrenic brain is not due to drugs and may be specific to one type of dopamine receptor. That it is the D2 receptor is supported by observations on the binding of flupenthixol, one component of which (attributed to the D2 receptor) is displaceable by the ligand domperidone, while the residual specific binding is assumed to be to the D1 receptor. Only the D2 component has been found to be significantly increased in schizophrenic brains (Table 6).

TABLE 6   Domperidone Displacement of ^{3}H-flupenthixol (2 nmol ligand concentration) binding in controls and drug-free schizophrenics (Cross, Crow and Owen, 1980).   mean ± s.e.m.

	n	D1 Component	D2 Component
Controls	9	104 ± 16	57 ± 9
Schizophrenics	7	133 ± 11	109 ± 16
		NS	$p < 0.02$

(Results as fmol/mg protein)

Clinical Correlations

The investigations of spiroperidol binding have now been extended to a series of 53 brains with a casenote diagnosis of schizophrenia by comparison with 37 controls. Maximum binding was increased in 30 patients in whom the Feighner criteria for a diagnosis of schizophrenia were fulfilled (mean 21.9 ± 9.8 fmol/mg tissue) by comparison with controls (mean 12.3 ± 6.5; $p < .001$). The increase in binding was not clearly related to the presence or absence of nuclear symptoms or to the presence of catatonic features (Table 7) when these features were assessed from the case notes.

TABLE 7 ^{3}H-spiroperidol Binding in Post-mortem Brain.

	n	Bmax (± S.D.)
Controls	39	12.3 ± 6.5
Schizophrenics:		
nuclear syndrome present	20	22.5 ± 9.9
nuclear syndrome absent	9	23.6 ±12.1
catatonic syndrome present	14	21.3 ± 9.5

It has been suggested (Crow, 1980) that changes in the dopamine receptor are related specifically to the presence of positive symptoms (delusions, hallucinations, and thought disorder or the type I syndrome) and that these changes are unrelated to the development of negative symptoms, flattening of affect and poverty

of speech (the type II syndrome or the "defect state") which may
be much more closely related to structural brain changes demonstrated by computerised tomography (Johnstone et al., 1976).
According to this hypothesis (Table 8) the presence of the type
I syndrome predicts potential response to neuroleptic drugs
while the type II syndrome predicts poor long-term outcome.

TABLE 8   Two Syndromes in Schizophrenia

	Type I	Type II
Symptoms	Delusions, halluci- nations, thought disorder (positive symptoms)	Flattening of affect, poverty of speech (negative symptoms)
Potential for response to drugs	Good	Poor
Intellectual impairment	Absent	Sometimes present
Outcome	Reversible	(?) Irreversible
Pathology	Increased dopamine receptors.	Cell loss and structural changes in the brain.

This hypothesis is based mainly upon clinical findings
concerning the selectivity of drugs for certain symptoms
(Johnstone, Crow, Frith, Carney and Price, 1978) but it has
recently become possible to test the prediction concerning
dopamine receptors directly on post-mortem brain. The brains
of a series of 14 patients who had been assessed in life by the
authors have now been assessed for spiroperidol binding after
death. Increased binding is strongly correlated with positive
but not negative symptoms (Table 9) as previously predicted.

TABLE 9 ^{3}H-spiroperidol Binding in the Brains of Schizophrenic
Patients assessed clinically before death.

	n = 14 r	p
Bmax vs. postive symptoms	0.722	<0.01
Bmax vs. negative symptoms	-0.338	N.S.

The changes in binding have been found to be unrelated to the presence or absence of movement disorder (Table 10).

TABLE 10 ^{3}H-spiroperidol Binding in Patients Assessed Ante-mortem

Schizophrenic Patients	n	Bmax
with movement disorder	3	21.5 ± 14.2
without movement disorder	9	25.6 ± 11.8

Thus increased spiroperidol binding appears to be unrelated to movement disorder and prior drug treatment but may be related to the presence of positive symptoms (the type I syndrome).

Conclusions
      Thus of the six neurohumoural hypotheses of schizophrenia noted in Table 1, all but one appear to have been excluded (at least as applying to a large proportion of schizophrenic patients) by recent post-mortem studies. The evidence to date, however, rather strongly supports the hypothesis that there are increased numbers of dopamine receptors in the brains of many patients with schizophrenia. The increase appears to be selective for the D2 receptor and is at least in part unexplained by previous neuro-leptic medication. Data from pre-mortem clinical assessments suggests that this change is not related to movement disorder but may be associated with the presence of positive symptoms (the type I syndrome) as had previously been predicted from pharmacological and radiological studies of acute and chronic schizophrenic patients.

Acknowledgments
      We would like to thank those psychiatrists and pathologists who have been of great assistance to us in obtaining brain specimens, in particular Dr. J. Burston, Professor J.A.N. Corsellis, Dr. M.D. Harris, Professor J. Hume Adams, Dr. W.S. Killpack, Dr. D.R. Oppenhiemer, Dr. R. Ransom, and Doctors G. Slavin and A. Price. We would like to thank Mr. Peter Mantel for help in administration and Ms. A.J. Davies for her secretarial assistance.

REFERENCES

Bird, E.D., Crow, T.J., Iversen, L.L., Longden, A., Mackay, A.V.P., Riley, G.J. and Spokes, E.G. (1979). J. Physiol. Lond., <u>293</u> 26-37P.

Bowers, M.B. (1974). Central dopamine turnover in schizophrenic syndromes. Arch. Gen. Psychiatry, <u>31</u>, 50-4.

Cross, A.J., Crow, T.J., Glover, V., Lofthouse, R., Owen, F., and Riley, G.J. (1977). Monoamine oxidase activity in post-mortem brains of schizophrenics and controls. Brit. J. Clin. Pharmac., <u>4</u>, 719P.

Cross, A.J., Crow, T.J., and Owen, F. (1980). ^{3}H-cis flupenthixol (3H-FPT) binding in post-mortem brains of schizophrenics is evidence for a selective increase in dopamine D-2 receptors. 12th CINP Congress Abstracts.

Cross, A.J., Crow, T.J., Killpack, W.S., Longden, A., Owen, F., Riley, G.J. (1978). The activities of brain dopamine-β-hydroxylase and catechol-o-methyl transferase in schizophrenics and controls. Psychopharmacology, <u>59</u>, 117-21.

Crow, T.J. (1980). Molecular pathology of schizophrenia: more than one disease process? Brit. Med. J., <u>280</u>, 66-68.

Crow, T.J., Owen, F., Cross, A.J., Johnstone, E.C., Joseph, M.H., and Longden, A. (1980). The dopamine receptor as the site of primary disturbance in schizophrenia. In press in <u>Enzymes and Neurotransmitters in Mental Disease</u> (Ed. E. Usdin).

Crow, T.J., Deakin, J.F.W., Johnstone, E.C., Longden, A. (1976). Dopamine and schizophrenia. Lancet, <u>ii</u>, 563-6.

Crow, T.J. Mitchell, W.S. (1975). Subjective age in chronic schizophrenia: evidence for a sub-group of patients with defective learning capacity? Br. J. Psychiatry, <u>126</u>, 360-3.

Crow, T.J. and Stevens M. (1978). Age disorientation in chronic schizophrenia; the nature of the cognitive deficit. Br. J. Psychiatry, <u>133</u>, 137-42.

Gaddum, J.H. (1954). Drug antagonistic to 5-hydroxytryptamine. In: Wolstenholme G.W., ed. Ciba Foundation symposium on hypertension. Boston: Little Brown 75-7.

Johnstone, E.C., Crow, T.J., Frith, C.D., Carney, M.W.P., Price, J.S. (1978). Mechanism of the antipsychotic effect in the treatment of acute schizophrenia. Lancet, <u>i</u>, 848-51.

Johnstone, E.C., Crow, T.J., Frith, C.D., Stevens, M., Kreel, L., and Husband, J. (1978). The dementia of dementia praecox. Acta Psychiatr Scand, <u>57</u>, 305-24.

Joseph, M.H., Baker, H.F., Crow, T.J., Riley, G.J., Risby, D. (1979). Brain tryptophan metabolism in schizophrenia: a post-mortem study of metabolites on the serotonin and kynurenine pathways in schizophrenic and control subjects. Psychopharmacology <u>62</u>, 279-85.

Lee, T., Seeman, P., Tourtellotte, W.W., Farley, I.J., Hornykeiwicz, O. (1978). Binding of ^{3}H-apomorphine in schizophrenic brains. Nature <u>274</u>, 897-900.

Mackay, A.V.P., Doble, A., Bird, E.D., Spokes, E.G., Quik, M.,
    and Iversen, L.L. (1978). ^{3}H-spiperone binding in normal
    and schizophrenic post-mortem human brain. Life Sciences 23,
    527-532.

Muller, P., Seeman, P. (1977). Brain neurotransmitter receptors
    after long-term haloperidol: dopamine, acetylcholine,
    serotonin, α-noradrenergic and naloxone receptors. Life Sci
    21, 1751-8.
Murphy, D.L. and Wyatt, R.J. (1972). Reduced monoamine oxidase
    activity in blood platelets from schizophrenic patients.
    Nature 238, 225-6.
Owen, F., Cross, A.J., Crow, T.J., Longden, A., Poulter, M.,
    and Riley, G.J. (1978). Increased dopamine-receptor
    sensitivity in schizophrenia. Lancet ii, 223-5.
Owen, F., Cross, A.J., Crow, T.J., and Waddington, J.L. (1980).
    Is the increase in dopamine receptors in schizophrenia brain
    due to drugs? A study of neuroleptic treated non-schizo-
    phrenic brains. 12th CINP Congress Abstracts.
Owen, F., Cross, A.J., Poulter, M., and Waddington, J.L. (1979).
    Change in the characteristics of ^{3}H-spiperone binding to rat
    striatal membranes after acute chlorpromazine administration:
    effects of buffer washing of membranes. Life Sciences 25,
    385-390.
Owen, F., Cross, A.J., Crow, T.J., Lofthouse, R., and Poulter, M.
    Neurotransmitter receptors in brain in schizophrenia. Acta
    Psychiatr Scand (Suppl) (in press).
Randrup, A., Munkvad, I. (1967). Stereotyped activities produced
    by amphetamine in several animal species and man. Psycho-
    pharmacologia, 11, 300-10.
Roberts, E. (1972). An hypothesis suggesting that there is a
    defect in the GABA system in schizophrenia. Neurosciences
    Res. Prog. Bull. 10, 468-482.
Stein, L., Wise, C.D., (1971). Possible etiology of schizophrenia
    progressive damage to the noradrenergic reward system by
    6-hydroxydopamine. Science 171, 1032-6.

Dopamine Metabolism in Human Brain

M. SANDLER, VIVETTE GLOVER, M. A. REVELEY,
PAULINE LAX and G. REIN
*Bernhard Baron Memorial Research Laboratories and
Institute of Obstetrics and Gynaecology, Queen
Charlotte's Maternity Hospital, London W6 0XG, UK*

The vast literature on dopamine which now exists attests to
the pre-eminence it has achieved among the monoamines, since its
first recognition as a neurotransmitter in the late 1950's
(Carlsson, 1959). Much of our latter-day interest has been sustained by the finding that L-dopa administration to parkinsonian
patients, resulting in dopamine generation in the central nervous
system, gives rise to substantial therapeutic benefit (Cotzias
et al.,1967). More recently, attention has focussed on the possible role of the amine in schizophrenia (Randrup and Munkvad,
1968). Despite this preoccupation with dopamine, certain metabolic pathways connected with its generation or disposition have
still been underinvestigated, particularly in man. This paper
sets out to provide a brief evaluation of the importance of four
of these pathways in the human brain, L-aromatic amino acid decarboxylase, L-dopa transaminase, monoamine oxidase and phenolsulphotransferase.

L-AROMATIC AMINO ACID DECARBOXYLASE

The opening session of the Fourth International Catecholamine
Symposium (Usdin et al., 1979) devoted itself to the question of
whether this decarboxylase is truly present in human brain and
what factors cause its activity to vary (Sacks et al., 1979). The
problem facing the panel was that a number of reputable laboratories had been unable to detect such activity in human brain
tissue (e.g. Sacks,1961; Robins et al.,1967; Vogel et al.,1969) -
although measurable activity had been noted by others (e.g. Lloyd
and Hornykiewicz,1970a; Wyatt et al.,1976), with a decrease in
Parkinson's disease (Lloyd and Hornykiewicz,1970b). In his symposium presentation, Sacks reported that since his first paper on
this theme (Sacks,1961) he had subsequently been able to detect
substantial activity in the brains of subjects who had died an

anoxic death although the enzyme was hard to detect in other brains. Whether pulmonary agents are able to activate the decarboxylase (Sandler,1979), perhaps prostaglandins or related compounds released from the pulmonary vascular bed in response to a local triggering mechanism (Alabaster and Bakhle,1976), must await further research.

Amongst other problems which are still unsolved is why the therapeutic benefit conferred by L-dopa, which depends on intact decarboxylase activity for conversion of the amino acid to dopamine, is neutralised by high dosage of pyridoxine (Duvoisin et al.,1969), a known cofactor for the decarboxylase. Although a number of explanations have been put forward (see Sandler,1972), none is entirely satisfactory. One possibility which has recently suggested itself concerns the ability of γ-aminobutyric acid (GABA) to inhibit pyridoxal kinase, the enzyme converting pyridoxal to pyridoxal phosphate. This inhibition is achieved by the action of the GABA-pyridoxal Schiff base complex (Abercrombie and Martin,1980) which may provide a parallel model for the situation in parkinsonism. There are indications that other amine-Schiff base derivatives have a similar action. Thus, it seems possible that an L-dopa- or dopamine-pyridoxal Schiff base prevents the generation of pyridoxal phosphate by pyridoxal kinase and thus interferes with L-dopa decarboxylation.

DOPA TRANSAMINASE

Although this enzyme is well described in rat (Fonnum et al., 1964; Tangen et al.,1965) or guinea pig (Fonnum and Larsen,1965) brain, we have recently been quite unable to find evidence of its presence in human brain. Whether this pyridoxal phosphate enzyme, in common with L-aromatic acid decarboxylase, behaves anomalously in man by virtue of some peculiar property connected with the pyridoxal kinase system, must be the subject of future investigation.

MONOAMINE OXIDASE

Some ten years ago, we held the view in this laboratory (e.g. Youdim,1973) that a specific monoamine oxidase for dopamine might exist. Since that time, the subdivision by Johnston (1968) of MAO into A and B forms has been widely accepted as a working classification and we have recently had the opportunity of reassessing the mechanism of the oxidative deamination of dopamine. Dopamine is a substrate for both forms of the enzyme (Yang and Neff,1973) and its K_m is similar for each (Glover et al.,1980a). Both human placenta, which contains pure MAO A (Egashira,1976) and human platelets, a pure source of MAO B (Donnelly and Murphy,1977) can metabolise dopamine, although dopamine is relatively less active than the specific MAO A substrate, 5-hydroxytryptamine, with placental enzyme and more active than the specific MAO B substrate, phenylethylamine, with the platelet enzyme.

We have approached the question of an additional MAO for dopamine by studying 5-hydroxytryptamine, phenylethylamine and dopamine oxidation in different regions of the human brain (Glover et al.,1980a). There was no region where the total dopamine oxidation could not be accounted for by the joint contribution of MAO A and B. Furthermore, there was a high degree of correlation between the sensitivity of dopamine oxidation to inhibition by 10^{-6}M (-)-deprenyl (a selective MAO B inhibitor) and the A/B ratio present, as measured directly with 5-hydroxytryptamine and phenylethylamine ($r = 0.84$, $p < 0.001$).

Thus, there is no need to invoke the existence of a separate dopamine MAO in human brain. A certain amount of confusion has arisen from the fact that human brain contains greater activity of MAO B than MAO A (Glover et al.,1977) and that proportionally more of this enzyme form is present in areas of the brain with high dopamine concentrations (Glover et al.,1980a). Thus, deprenyl, in its optimal selective inhibitory concentration for human brain of 10^{-6}M, gives rise to a higher degree of inhibition of dopamine oxidation in such dopamine-rich areas as the caudate or nucleus accumbens than in frontal or cerebellar cortex. The selective action of deprenyl on dopamine oxidation in the striatum is thus sufficient to account for the benefit achieved by this inhibitor when used in combination with L-dopa in Parkinson's disease. It should be noted that although MAO B tends to predominate in human brain, the picture is different in rat brain which is richer in MAO A (Waldmeier et al.,1976).

There have been many claims that platelet MAO activity is decreased in schizophrenia (for review,see Wyatt et al.,1979) and a particular decrease has been noted in the paranoid form of the disease (Schildkraut et al.,1976). Many laboratories are unable to confirm these findings. Even so, it is possible that the discrepancy derives from differences in the population under investigation and/or variations in technique. Because platelet MAO consists wholly of MAO B, as noted above, and because this enzyme form predominates in human brain, a number of authors have sought evidence of its reduced activity in this tissue (see Reveley et al.,1980). However, in the various small series published to date, no evidence of such a decrease was evident. Even the two larger studies (Schwartz et al.,1974; Crow et al.,1979) only included nine schizophrenic brains each. We have now completed an investigation (Reveley et al.,1980) of MAO activity from ten well-diagnosed schizophrenics, ten schizophrenia-like psychotics and 20 controls, all supplied by Dr.E.D.Bird. Two separate areas were sampled from each brain. There was no significant difference in enzyme activity between psychotics and controls, using either 5-hydroxytryptamine, dopamine or phenylethylamine as substrates. This finding was apparent whether material from the patient group was evaluated collectively or divided into its two constituent sub-groups, diagnosed schizophrenic or schizophrenia-like illness.

PHENOLSULPHOTRANSFERASE

This enzyme, which transfers the sulphate moiety from "active sulphate" (PAPS) to an acceptor amine has also been relatively little studied. However, recent investigations (Rein et al.,1980) indicate that its action is likely to be of more than academic interest - the K_m of dopamine for the enzyme, 3 μM, is considerably lower than that of dopamine for either MAO A or MAO B. Thus, at low substrate concentrations, phenolsulphotransferase will be more important relative to MAO than at high levels.

Very recently, we and other groups (Hart et al.,1979; Anderson and Weinshilboum,1980; Rein et al.,1980) have all been able to detect substantial activity of this enzyme in the human platelet. It thus seemed important to determine whether the platelet enzyme is typical of that present in human brain and elsewhere in the body. Preliminary studies immediately indicated that it manifests a different type of substrate preference from the enzyme of rat liver. However, we have been able to show that the activity present in human jejunal biopsy material and fresh placenta and, indeed, in human brain (Renskers et al.,1980) looks very similar to that in the platelet (Rein et al.,1980). In common with the recent findings of Renskers et al. (1980) we have so far only been able to detect low activity of this enzyme in the postmortem human brain (in preparation). However we know that phenolsulphotransferase is rather unstable, and we will need brain biopsy material to evaluate the relative activity in brain and other tissues.

As the human platelet and brain enzyme appear to have similar properties, the accessibility of the platelet and the relative simplicity of the assay procedure will enable further studies to be carried out in neuropsychiatric disease. Very recently, we have been able to demonstrate greater activity of platelet phenolsulphotransferase activity in parkinsonian subjects than in normal controls (Glover et al.,1980b). However, this phenomenon seems likely to represent an adaptive increase to L-dopa, for when the platelets of parkinsonian patients untreated by L-dopa were examined, they failed to show this rise in enzyme activity (in preparation). Thus, it seems likely that we have demonstrated an adaptive increase of enzyme resulting from a drug effect. It is not unreasonable to suppose that an adaptive increase of phenolsulphotransferase takes place in human brain similarly and, potentially at least, may siphon off transmitter dopamine from areas where it can be of most usefulness. Thus, a quite new approach to the treatment of Parkinson's disease is opened up, the administration of a phenolsulphotransferase inhibitor, as an adjuvant to other available therapeutic L-dopa combinations.

REFERENCES

Abercrombie,D.M. and Martin,D.L. (1980). Inhibition of pyridoxal
kinase by the pyridoxal-γ-aminobutyrate imine. J.biol.Chem.,
255, 79-84.

Alabaster,V.A. and Bakhle,Y.S. (1976). Release of smooth muscle-
contracting substances from isolated perfused lungs. Eur.J.
Pharmac., 35, 349-360.

Anderson,R.J. and Weinshilboum,R.M. (1979). Phenolsulphotransferase: enzyme activity and endogenous inhibitors in the human
erythrocyte. J.Lab.Clin.Med., 94, 158-171.

Carlsson,A. (1959). The occurrence, distribution and physiological role of catecholamines in the nervous system. Pharmacol.
Rev., 11, 490-493.

Cotzias,G.C., van Woert,M.H. and Schiffer,L.M. (1967). Aromatic
amino acids and modification of Parkinsonism. New Engl.J.Med.,
276, 374-379.

Crow,T.J., Baker,H.F.,Cross,A.J.,Joseph,M.H.,Lofthouse,R.,Long-
den,A.,Owen,F.,Riley,G.J.,Glover,V. and Killpack,W.S. (1979).
Monoamine mechanisms in chronic schizophrenia: postmortem
neurochemical findings. Br.J.Psychiat., 134, 249-256.

Donnelly,C.H. and Murphy,D.L.(1977). Substrate and inhibitor-
related characteristics of human platelet monoamine oxidase.
Biochem.Pharmac., 26, 853-858.

Duvoisin,R.C.,Yahr,M.D. and Cote,L.D. (1969). Pyridoxine reversal
of L-dopa effects in Parkinsonism. Trans.Amer.neurol.Ass.,
94, 81-82.

Egashira,T. (1976). Studies on monoamine oxidase. XVIII. Enzymic
properties of placental monoamine oxidase. Jap.J.Pharmacol.,
26, 493-500.

Fonnum,F.,Haavaldsen,R. and Tangen,O. (1964). Transamination of
aromatic amino acids in rat brain. J.Neurochem., 11, 109-118.

Fonnum,F. and Larsen,K. (1965). Purification and properties of
dihydroxyphenylalanine transaminase from guinea pig brain.
J.Neurochem., 12, 589-598.

Glover,V.,Elsworth,J.D. and Sandler,M. (1980a). Dopamine oxidation and its inhibition by (-)-deprenyl in man. J.neural
Transmiss., Suppl.16, 163-172.

Glover,V.,Sandler,M.,Owen,F, and Riley,G.J.(1977). Dopamine is a
monoamine oxidase B substrate in man. Nature,Lond., 265,
80-81.

Glover,V.,Sandler,M.,Rein,G.,Ward,C. and Stern,G. (1980b).Mono-
amine oxidase and phenolsulphotransferase in Parkinson's disease, In Progress in Parkinson's Disease, (eds. F.Clifford
Rose and R.Capildeo), Pitman Medical, London, in press.

Hart,R.F.,Renskers,K.J.,Nelson,E.B. and Roth,J.A. (1979). Localization and characterization of phenol sulfotransferase in
human platelets. Life Sci., 24, 125-130.

Johnston,J.P. (1968). Some observations upon a new inhibitor of
monoamine oxidase in brain tissue. Biochem.Pharmac., 17,
1285-1297.

Lloyd,K. and Hornykiewicz,O. (1970a). Occurrence and distribution
of L-DOPA decarboxylase in the human brain. Brain Res., 22,
426-428.

Lloyd,K. and Hornykiewicz,O. (1970b). Parkinson's disease: activity of L-dopa decarboxylase in discrete brain regions.
Science,170, 1212-1213.

Randrup,A. and Munkvad,I. (1968). Behavioural stereotypes induced
by pharmacological agents. Pharmakopsychiat.Neuro-Psychopharmak.,
1, 18-26.

Rein,G., Glover,V.and Sandler,M. Sulphate conjugation of biologically active monoamines and their metabolites by human platelet
phenolsulphotransferase. Submitted for publication.

Renskers,K.J.,Feor,K.D. and Roth,J.A.(1980). Sulfation of dopamine
and other biogenic amines by human brain phenol sulfotransferase. J.Neurochem., 34, 1362-1368.

Reveley,M.A.,Glover,V.,Sandler,M. and Spokes,E.G. (1980). Brain
monoamine oxidase activity in schizophrenics and controls:
relationship to diagnosis, sex, and age. Submitted for publication.

Robins,E.,Robins,J.M.,Croniger,A.B.,Moses,S.G.,Spencer,S.J. and
Hudgens,R.W.(1967). The low level of 5-hydroxytryptophan
decarboxylase in human brain. Biochem.Med., 1, 240-251.

Sacks,W. (1961). A cerebral decarboxylase for 5-hydroxytryptophane
in humans. J.Appl.Physiol., 16, 1050-1054.

Sacks,W., Vogel,W.H.,Nagatsu,T.,Lloyd,K.G. and Sandler,M.(1979). Is there DOPA decarboxylase in human brain? In Catecholamines: Basic and Clinical Frontiers, (eds.E.Usdin,I.J.Kopin and J.Barchas), Pergamon Press, New York, pp.127-131.

Sandler,M. (1972). Catecholamine synthesis and metabolism in man (with special reference to parkinsonism). In Handbook of Experimental Pharmacology, Vol.33, Catecholamines, (eds.H.Blaschko and E.Muscholl), Springer, Berlin, pp.845-899.

Sandler,M. (1979). Is there dopa decarboxylase in human brain? In Catecholamines: Basic and Clinical Frontiers, (eds.E.Usdin, I.J.Kopin and J.Barchas), Pergamon Press, New York,pp.130-131.

Schildkraut,J.J.,Herzog,J.M.,Orsulak,P.J.,Edelman,S.E.,Shein, H.M. and Frazier,S.H.(1976). Reduced platelet monoamine oxidase activity in a subgroup of schizophrenia patients. Am.J. Psychiat., 133, 438-440.

Schwartz,M.A.,Aikens,A.M. and Wyatt,R.J. (1974). Monoamine oxidase activity in brains from schizophrenics and mentally normal individuals. Psychopharmacologia, 38, 319-328.

Tangen,O.,Fonnum,F. and Haavaldsen,R. (1965). Separation and purification of aromatic amino acid transaminases from rat brain. Biochim.Biophys.Acta, 96, 82-90.

Usdin,E.,Kopin,I.J. and Barchas,J. (1979). Eds.Catecholamines: Basic and Clinical Frontiers,Pergamon Press, New York.

Vogel,W.H.,Orfei,V. and Century,B. (1969). Activities of enzymes involved in the formation and destruction of biogenic amines in various areas of human brain. J.Pharmacol.exp.Ther., 165, 196-203.

Waldmeier,P.C.,Delini-Stula,A. and Maitre,L.(1976). Preferential deamination of dopamine by an A type monoamine oxidase in rat brain. Naunyn-Schmiedeberg Arch.Pharmac., 292, 9-14.

Wyatt,R.J.,Erdelyi,E.,Schwartz,H.,Herman,M. and Barchas,J.D. (1978). Difficulties in comparing catecholamine-related enzymes from the brains of schizophrenics and controls. Biol.Psychiat., 13, 317-334.

Wyatt,R.J.,Potkin,S.G. and Murphy,D.L.(1979). Platelet monoamine oxidase activity in schizophrenia: a review of the data. Am.J.Psychiat., 136, 377-385.

Yang,H-Y.T. and Neff,N.H. (1974). The monoamine oxidases of brain: selective inhibition with drugs and the consequences for the metabolism of biogenic amines. J.Pharmac.exp.Ther.189,733-740.

Youdim,M.B.H. (1973). Multiple forms of mitochondrial monoamine oxidase. Br.Med.Bull., <u>29</u>, 120-122.

Propranolol Binding in Human Brain: Preliminary Studies

GAVIN P. REYNOLDS*, PETER RIEDERER* and EBERHARD GABRIELt

* Neurochemistry Group, Ludwig Boltzmann Institute
  for Clinical.Neurobiology, Lainz Hospital, A-1130
  Vienna, Austria
t Psychiatric Hospital, Baumgartner Höhe, A-1140
  Vienna, Austria

INTRODUCTION

It has been established by several groups in the
past few years that large dosage of (+)propranolol
can be effective in the treatment of schizophrenia
and other psychiatric disorders including anxiety
(Roberts and Amacher, 1978). Whether this effect
is due to a central ß-adrenergic receptor blockade
is unclear. There are reports that the (+)isomer of
propranolol, much less active as a ß-blocker, is as
effective an antipsychotic as (-) propranolol although
the observation that the antipsychotic dose of (+)pro-
pranolol concides with that for optimal peripheral
ß-blockade (quoted by Grüter, 1977) appears inconsistent. Other postulated mechanisms for this apparently
central action include blockade of serotonin (5-HT)
receptors (Green and Grahame-Smith, 1976), monoamine
oxidase inhibition (Milmore and Taylor, 1976) and
membrane stabilisation (Bainbridge and Greenwood,1971).

In order to shed light on this effect, we have carried
out studies in vitro investigating the action of
propranolol using post-mortem human brain samples.
In particular we have characterized the specific
binding of radiolabelled (+)propranolol to ß-adrenergic
receptors in the human hippocampus, and compared
samples isolated from schizophrenics with those from
control patients.

METHODS

^{3}H($\pm$)propranolol binding was a modification of the
method of Nahorsky (1976), adapted to the brain homogenate pretreatment used here. Brain samples, dissected conservatively to ensure that no parts of other
brain regions were present, were homogenised using
a Teflon-glass Potter homogenizer with nine volumnes
of 10mM tris buffer (pH 7,5) containing 0.1%
ascorbate. This was stored frozen at -40C until
analysis, when the homogenate was diluted ten-fold
with 50mM tris buffer (pH 8.1) containing 15mM
MgCl and 0.1% ascorbate and centrifuged at 50 000g
for 15 min. After decanting the supernatant and
reconstitution in an equal amount of this buffer,
this step was repeated. Triplicate 1 ml samples of the
final 1% preparation was incubated for 30 min at
25C with tritiated ($\pm$)propranolol in the presence or
absence of 0.2mM isoproterenol. After addition of
5 ml buffer at 4C and vacuum filtration through
Whatman GF/A filters with two washes of 5 ml buffer,
the amount of label bound to the brain homogenate on
the filter was determined by scintillation counting.

Spiperone binding was measured using the method of
Reisine et al. (1977), and monoamine oxidase (MAO)
as published by Tipton and Youdim (1976).

RESULTS AND DISCUSSION

We decided to use propranolol in these studies in
order to further our understanding of its action in
psychiatric medicine. Unfortunately it is not the
best ligand for characterisation of ß-receptors since
it has a high lipophilicity and hence a high non -
specific binding component. While this could have
been decreased by fractionating the brain homogenate
and working with a purified membrane fraction, we
decided to use a simple work-up procedure in order
to minimise any possible artifactual effects.

The ß-receptor (isoproterenol-specific) binding of
propranolol to human brain preparations was character-
ised by Scatchard analysis. A typical example is
shown in figure 1 where a range of propranolol concentrations between 0.5 and 31nM have been used. There

106

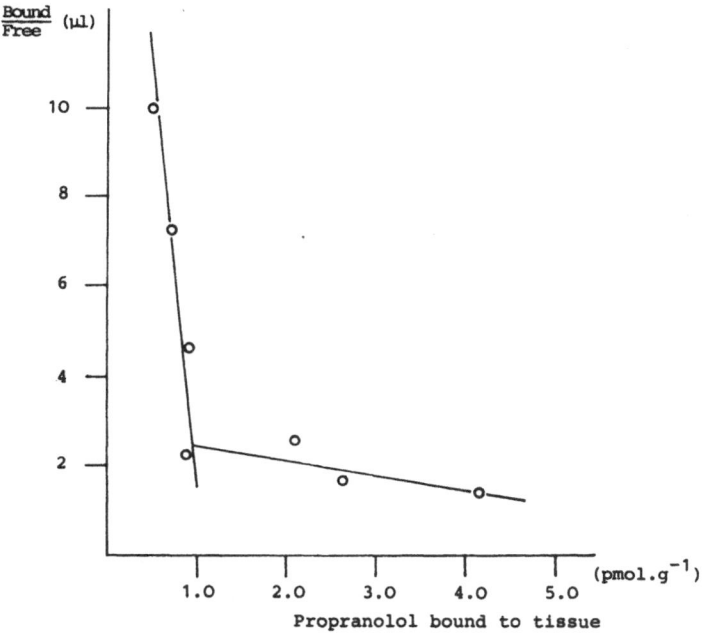

Figure 1   Scatchard analysis of propranolol
             binding to human hippocampus

appear to be two components of this specific binding,
one with high affinity and low capacity, one with a
lower affinity but higher capacity.

We have studied thsi high affinity binding by
Scatchard analysis using (+) propranolol concentrations of 0.25 to 4 nM. The hippocampus was chosen for
study since this area, as part of the limbic system,
has been implicated in the antischizophrenic action
of propranolol (Gruzelier et al., 1978). In a series
of 7 hippocampus preparations, taken post—mortem from
patients who had not exhibited any neuropsychiatric
disease, mean values were as follows: Apparent
K_D = 0.96 nM (s.e.m. 0.14 nm), B_{max} = 1.75 pmol.g^{-1}
(s.e.m. 0.12 pmol.g^{-1}). No correlation with age or
sex was observed. The specific binding at 4nM (which
should approximate to maximal binding for the high
affinity receptor) was found to be totally inhibited
by 10^{-6}M noradrenaline, yet unaffected by 5-HT;

107

this being further evidence for the high affinity
binding being ß-adrenergic.

This binding has also been studied by Scatchard
analysis in samples of hippocampus taken post mortem
from patients diagnosed as schizophrenic.
Results, including some relevant information on these
patients, are shown in table 1.
All patients, with the possible exception of No. 4,
had met both the ICD classification and Feighner's
research criteria for schizophrenia, and all had
paranoid components.

In order to minimise spurious post-mortem effects,
the brains investigated were chosen for short intervals between death and freezing, which in all cases
was less than 12 hours.

From table 1 there appears no overall significant
difference from normal in the ß-noradrenergic binding
of propranolol to schizophrenic hippocampus. K_D values
do, however, tend to be higher. That there may be a
genuine difference is under further investigation:
drug effects may well play a role since antidepressant
medication is known to decrease ß-receptor sensitivity
(Banerjee et al., 1977). In this respect it is notable
that the two patients who had received tranquilizer
medication (Nos. 1 and 2) exhibit the lowest levels
of maximum binding.

Since these studies have been carried out using the
racemic form of propranolol, it is important to confirm that this supposedly ß-receptor binding is
stereospecific. And indeed it is: the less active
(+) propranolol inhibits the high affinity binding,
with an IC_{50} of approximately 100 nM. Thus the
apparent K_D values quoted here derive exclusively

TABLE 1

Propranolol binding to hippocampus preparations from Schizophrenics

Patient	Age	Sex	K_D (nM)	B_{max} (pmol.g tissue^{-1})	Medication NL	Park.	Tranq.	Last NL medication: time before death
1	67	f	0.76	1.28	+	−	+	29 d
2	59	m	1.36	1.13	+	+	+	2 d
3	78	f	1.51	1.78	+	−	−	3 d
4	73	f	1.62	1.62	+	+	−	40 d (depot)
5	64	f	1.22	1.99	−	−	−	none
mean	68.2		1.29	1.56				
s.e.m.	3.3		0.15	0.16				
controls (n = 7):								
mean	68.7	(5 f 2 m)	0.96	1.75				
s.e.m.	6.5		0.14	0.20				

post mortem time < 10 h

NL   = neuroleptic
Park = antiparkinsonian
Tranq = tranquilizer
+ = present
− = absent

109

from (-)propranolol and are thus two-fold too high: the K_D for (-) propranolol binding is therefore 0.48 nM.

The low affinity, higher capacity binding observed in figure 1 poses an interesting problem. It may be that this is similar to the lower affinity mitochondrial binding of dihydroalprenolol observed by Williams and Lefkowitz (1978); further studies with cell fraction preparations should clarify this. It is unlikely that these are α-receptors (although the isoproterenol concentration used could well be high enough to bind to these sites) since their K_D for propranolol (in calf cerebellum) is much higher than the concentrations used here (U'Prichard and Snyder, 1977).

It is tempting to speculate, however, that the effect observed here represents the detection of "uncoupled", desensitized ß-receptors (discussed in Williams and Lefkowitz, 1978). Unfortunately we have not yet fully characterized this binding since at the high ligand concentrations required, variation in the unspecific binding tends to increase the inaccuracy of the results. However preliminary studies indicate that the binding is less sensitive to NA and is, to some extent, stereospecific.

To complement these studies we have measured two further central effects of the propranolol isomers: MAO inhibition and dopamine receptor blockade. Table 2 shows no great difference between the isomers in the inhibition of 5-HT oxidation; the results are similar to those from rat brain (Milmore and Taylor, 1976). Notable is that the effect is specific for A-type MAO; phenylethylamine oxidation is less affected.

TABLE 2

IC_{50} for the inhibition of MAO from human striatum by propranolol isomers

	Substrate	
	PE	5-HT
(+)propranolol	$>10^{-3}$	2×10^{-4}
(-)propranolol	$>10^{-3}$	4×10^{-4}

110

This inhibition of spiperone binding to the (dopamine) neuroleptic receptor shows a very similar IC_{50} of 3×10^{-4} M for both propranolol isomers. Thus these two very different effects may have the same origin, perhaps in a general disturbance of membrane function. If a membrane effect is contributory to the action of propranolol in psychiatric medicine, then it is likely to be multifactorial, affecting many such membrane-bound enzymes and receptors.

ACKNOWLEDGEMENTS

We thank Prof.K.Jellinger and his staff for dissection, the European Science Foundation and Hoechst Austria for financial support, I.C.I.Pharmaceuticals for propranolol isomers, Veronika Schay for excellent technical assistance and Inge Riederer for typing the manuscript.

REFERENCES

Bainbridge, J.G., and Greenwood, D.T. (1971) Neuropharmacol.,10, 453-458.

Banerjee, S.P., Kuang, L.S.,Riggi,S.J. and Chanda,S.K. (1977). Development of ß-adrenergic receptor subsensitivity by antidepressants. Nature, 268, 455-456.

Green, A.R. and Grahame-Smith, D.G. (1976). (-)Propranolol inhibits the behavioural responses of rats to increased 5-hydroxytryptamine in the central nervous system. Nature, 262, 594-596.

Grüter, W. (1977). Psychoses and Beta-blockers. In Beta-blockers and the central nervous system (ed.P.Kielholz),Hans Huber, Bern pp 235-251.

Gruzelier, J.H., Conolly, J.F., Yorkston, N.J., Zaki, S.A. and Hirsch S.R. (1978). Modulation of the orienting reaction and its habituation in schizophrenic patients treated with propranolol. In Abstracts of the 11th C.I.N.P. Congress, Vienna.

Milmore, J.E. and Taylor K.M. (1976). Propranolol inhibits rat brain monoamine oxidase. Life Sci.,17, 1843 - 1848

Nahorsky, S.R. (1976). Association of high affinity stereospecific binding of [^{3}H]propranolol to cerebral membranes with beta-adrenoceptors. Nature, <u>259</u>, 488 - 489.

Reisine, T.D., Fields,J.Z., Yamamura, H.I., Bird,E.D., Spokes, E., Schreiner, P.S. and Enna, S.J. (1977). Neurotransmitter receptor alterations in Parkinson's disease. Life Sci., <u>21</u>, 335-344.

Roberts, E. and Amacher, P., eds. (1978). <u>Propranolol and Schizophrenia</u>, Alan R. Liss, New York.

Tipton, K. and Youdim, M.B.H. (1976). Assay of mono-amine oxidase. In <u>Monoamine oxidase and its inhibition</u> (eds. Wolstenholme G.E.W. and Knight, J.) Elsevier, Amsterdam pp 393-403.

U'Prichard, D.C. and Snyder, S.H. (1977). Differential labelling of α and ß-noradrenergic receptors in calf cerebellum membranes with ^{3}H-adrenaline. Nature,<u>270</u>, 261-262.

Williams, L.T. and Lefkowitz, R.J. (1978). <u>Receptor binding studies in adrenergic pharmacology</u>, Raven Press, New York.

Section IV
Low – Molecular – Weight
Compounds

Dopamine and Noradrenaline in the Magno- and Parvocellular Portions of the Red Nucleus

K. JELLINGER, G. P. REYNOLDS and P. RIEDERER

Ludwig Boltzmann-Institute for Clinical Neurobiology,
Lainz Hospital, Wolkersbergenstrasse 1, A-1130 Vienna,
Austria

INTRODUCTION

The presence of dopamine (DA) in the red nucleus (RN) of the human has been reported by several laboratories (Sano,1959; Ehringer and Hornykiewicz,1960; Bertler, 1961; Birkmayer et al.,1974; Moses and Robins,1975; Jellinger and Riederer, 1977; Hornykiewicz,this volume). In these studies the DA concentration ranges from 0,19 to 1,17 µg/g tissue. Several factors presumably contribute to this substantial variation. Firstly, while differences in analytical methodology should not, in theory, contribute to such effects, in practice different methods often give different results as they reflect different levels of specificity. Thus it is relevant that the highest value is also the earliest reported, when the method was limited to a simple spectrophotometric procedure. Secondly, the results may reflect various differences in the sample tissue, such as a different patient population (with relevant factors including age, hospitalization, sex, time of death, race, etc.) or tissue history (death - to autopsy time and temperature, storage time, etc.) (for example, see Riederer and Wuketich,1976; Adolfsson et al.,1979). A further factor, that of differences in the brain region dissected, can be avoided by precise and unequivocal definition of the dissected tissue. In response to the discrepancies reported for the DA content of the human RN (Hornykiewicz, this volume) we have investigated the intra-regional distribution of DA and noradrenaline (NA) in this brain area.

MATERIAL AND METHODS

CLINICAL MATERIAL

Human brainswere obtained at autopsy and immediately
processed as previously described by Wuketich et al.
(1980). Analyses were performed on brain material from
a total of 10 cases. There were 6 males and 4 females
with ages ranging from 66 to 90 years, the average
age being $80,4 - 2,1$ years (mean $-$ s.e.m.). All cases
were free of neurologic disorders; there were no brain
injuries or cerebral infarctions, and none died in
long-lasting coma. For further clinical detail see
table 1.
Drug treatment included antibiotics, heart drugs and
diuretics. The study excluded patients on medications
that affect catecholamine metabolism (neuroleptics,
dopa, etc.). The time interval between death and brain
dissection ranged from 3 to 9 hours, average time
interval being $6,5 - 0,7$ hours (mean$\pm$s.e.m.).

DISSECTION

For the purpose of dissection, fresh (unfrozen) brains
were used which were placed on a plate cooled at -10
to -15° C. Dissection was performed using a dissecting knife and curved scissors. A deep cut was made
through the midbrain at the level of the superior
colliculus in a plane at right angles to the long axis
of the brain stem, comparable to fig. 45 of the atlas
of DeArmond et al. (1976) and fig. 67 of the atlas of
Nieuwenhuis et al. (1979). This section divides the
RN between its middle and lower two thirds (fig.1).
Parallel sections were made at the upper level of
the mid-brain above the RN, and at the level of the
inferior colliculus, and the brain slices were placed
on a glass plate kept on a cooled plate as above.
The discrete areas of the RN were isolated with
reference to the above atlases using a LEDU 271
magnifier lamp with five-fold magnification.
First, the magnocellular (MC) part of the RN located
in the medioventral part of the caudal third of the
nucleus, dorsal to the oculomotor nerve and lateral
of the interpeduncular nucleus was dissected. Then
the remaining gray areas of the RN of both the
upper and lower brainstem slices were dissected,
avoiding an inclusion of adjacent parts of myelinated
areas. Finally, the surrounding myelinated capsule
and fibers were dissected from both the posterior and

anterior, oral and caudal parts of the area.
This fraction includes various fiber tracts surrounding
the RN, e.g. rubrospinal tract, perirubral and
cerebellorubral tract, and components of central
tegmental tract which were not separated from each
other due to scarcity of material.

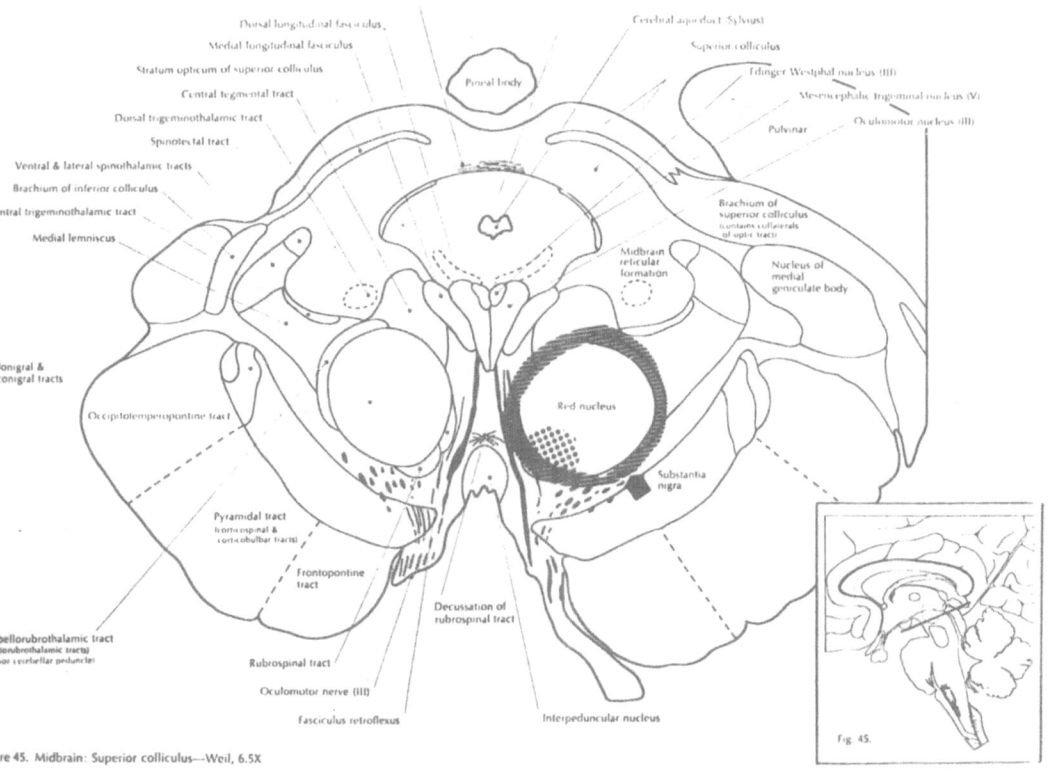

Fig.1 Horizontal section of the midbrain at the
       level of the superior colliculus (fig.45 of
       the atlas of DeArmond et al.,1976). RN with
       magnocellular part (pointed), parvocellular
       part (white), and white matter capsule (lined
       circle). Note vicinity of anterior parts of
       "capsule" to substantia nigra (arrow)

117

TABLE 1. Characterization of the Human Brain Tissue and Clinical Data of the Patients

Patient	sex	age yrs	post-mortem time (hrs.)	diagnosis (cause of death), post-mortem findings
W.A.	m	86	6	bronchopneumonia; senile brain
N.R.	m	80	5	vascular encephalopathy, status lacunaris of basal ganglia; pneumonia
R.A.	f	86	9	myocardial infarction, pulmonary edema; normal brain for age
M.H.	m	85	6	senile dementia; senile brain; pneumonia
B.M.	f	80	3	mammary carcinoma; brain normal for age
B.F.	m	80	4	chronic myelogenous leukemia; brain normal for age
B.R.	m	66	5	myocardial infarction, normal brain
L.K.	f	77	9	liver carcinoma; pulmonary metastases; status lacunaris of basal ganglia
B.A.	f	90	9	pulmonary embolism; senile brain
G.A.	m	74	9	old cerebral infarction in area of right middle cerebral artery

The procedure determination of catecholamines in
brain tissue involved an extraction based on the
method of Keller et al. (1976) followed by high
pressure liquid chromatographic (HPLC) separation
using the system described by Hjemdahl et al. (1979).

EXTRACTION

Up to 0,2 g of tissue was homogenized ultrasonically
in 1 ml of 0,1 M perchloric acid with 200 ng
3,4-dihydroxybenzylamine internal standard. This was
centrifuged (10.000 g, 5 min.) and 200 μl of the
supernatant was added to approximately 20 mg activated
alumina and 1 ml 0,5 M Tris buffer(pH 8,6) containing
0,04 M NaHSO$_3$ and 2 mM EDTA. After mixing for 15 min.,
the alumina was washed three times with water containing 1 % tris buffer. The catecholamines were desorbed
into 100 μl 0,4 M perchloric acid.

CHROMATOGRAPHY

50 μl of this extract was injected into the HPLC
column (Nucleosil 10SA) at 1 ml min^{-1} of pH 5,2
acetate citrate buffer (Keller et al.,1976). An
electrochemical detector system (LC-2A Bioanalytical
systems) set at an oxidizing potential of 0,6 V
enabled the amines to be determined. Quantification
was effected by measurement of chromatogram peak
heights relative to the internal standard and comparison with external standard mixtures.

RESULTS AND DISCUSSION

In an attempt to explain the large variability in the
DA concentration of the RN, this nucleus has been
divided into its magnocellular (MC) and parvocellular
(PC) part. Moreover, the DA concentration of the
surrounding tissue (white matter capsule) has been
measured. Furthermore, high pressure liquid chromatography (HPLC) with electrochemical detection has
been used for the measurement of catecholamines to
rule out a possibly unspecific fluorescence (and
hence too high concentrations of DA) of this latter
method.

Table 2 shows that there is a large difference in the
concentration of the PC and the MC parts of the human

TABLE 2. Catecholamine Concentration (ng/g fresh tissue) in Various Parts of the Human Red Nucleus

Controls (3)	R E D N U C L E U S		
	parvocellular portion	magnocellular portion	surrounding white matter capsule
DOPAMINE	163,9 $\pm$ 69,8	913,3 $\pm$ 165,5	976,5 $\pm$ 411
NORADRENALINE	379,2 $\pm$ 102,3	493,6 $\pm$ 150,6	600,7 $\pm$ 346,4

means $\pm$ s.d.

1) see table 1 for clinical details.
Analyses of red nucleus preparations were performed within one week after dissection (storage temperature: -40°C).

RN. This difference is highly significant for DA.
Noradrenaline (NA) tends to be higher in the MC part
but this difference is not significant in our limited
material.

The surrounding "white matter" capsule contains quite
large concentrations of both catecholamines (table 2).
These findings might explain, at least in part, the
large discrepancies of DA concentrations reported
in the literature because of the possibility that
dissected tissue was composed of both portions of the
RN. However in most studies no information is given as to
how the RN tissue has been dissected. Although the
RN is a well-defined anatomical structure and easy
to dissect, it seems to be of importance to describe
exactly from which anatomical part of the nucleus the
tissue has been taken. This is of special interest
because of the fact that quantitative dissection of
nuclei, sub-nuclei, etc. has not been a frequently used procedure,
but it might help in obtaining more consistent results (Wuketich
et al., 1980).

In addition to the data of table 2, which were obtained in material analysed within one week after
autopsy and dissection, RN tissue was analysed which
had been stored at -30°C between two weeks and two
years. In this conservatively dissected parts of the
RN, particularly containing the PC part, a DA concentration of 87,7 $\mp$ 21,4 (n=7) ng/g fresh weight was
found, whereas the NA concentration was 423,2$\mp$125,3
(n=7) ng/g fresh tissue. From this limited material it
can be suggested that storage time has some negative influence on DA concentration, which has not
been observed to such extent in other regions such as
caudate nucleus or putamen (Riederer and Wuketich,1976).
Therefore, much care should be taken in future studies showing regional dependencies of storage time
with the concentration of biogenic amines.

A dependence of DA or NA in RN from age could not be
determined in this preliminary study. It should be
noted, however, that our patients had a mean age of
80,4 years (table 1).

In agreement with the biochemically identified DA
in the RN (for references see introduction and data
of table 2) a weak activity of tyrosine hydroxylase
(TH) (Riederer et al.,1978), and monoamine oxidase
(Birkmayer et al.,1976) has been found. Further

evidence that DA is present in the human RN comes from
data showing that homovanillic acid, a main metabolite
of DA degradation could be detected in this region
(Bernheimer,1964, Hornykiewicz et al.,1968).

Histochemistry and electron microscopy of the RN have
shown characteristic changes in the composition of
this area during phylogenesis. For example, the RN
of the turtle's brain (Testudo hermanni) consists
mainly of a MC part with a dense catecholaminergic
fluorescence (Ovtscharoff 1972a). During phylogenesis
the pars magnocellularis becomes smaller, whereas
the PC portion develops to the main part in mammals.
This corresponds with a diminuation of the catecholaminergic activity (Ovtscharoff,1972b). Although,
histochemical analyses of the RN of mammals do not
confirm the biochemical data (Dahlström and Fuxe,1964;
Poitras and Parent,1978; Shimada et al.,1976)
a detailed study of the MC part of the RN is lacking.
Recent studies, however, using TH-immunohistofluorescence techniques show TH staining in the MC portion of the RN (J. Pearson,
personal communication).

The major descending pathway of the RN originates from
the MC part of the RN. This rubrospinal tract might
have some importance in the light of recent findings
regarding a dopaminergic innervation of the spinal
cord. DAergic neurons demonstrated in the
rat spinal cord have been shown to be independent of
NA neurons and are suggested to be parent in the
terminal of descending axons, particularly in the
sympathetic lateral and medial columns (Commissiong
et al.,1978). The highest concentration of DA is
found in the dorsal horn and in the zona intermedia,
part of which is considered to occur in cell bodies
of descending fibres originating in the substantia
nigra (Commissiong and Neff,1979). However,
destruction of the substantia nigra by 6-hydroxy-DA
caused only a 50 % reduction of DA in the ipsilateral
spinal cord. Therefore one might speculate some other
origin of these DA terminals in the spinal cord.
One possible source might be the MC part of the RN
since in mammals these fibres are known to end in
Rexed's laminae V, VI and the dorsolateral part
of laminae VII (see Massion,1967; Miller and Strominger,1973). Part of these sites of termination may
correspond to the lateral sympathetic areas of the
spinal column containing DAergic neuron (Commissiong

and Neff,1979). However, neuroanatomical and biochemical evidence for the DAergic innervation of the rubrospinal tract remains to be determined.

REFERENCES

Adolfsson, R., Gottfries, C.G., Roos, B.E., Winblad, B. (1979). Post-mortem distribution of dopamine and homovanillic acid in human brain, variation related to age, and a review of the literature. J.Neural Transm. 45, 81-105.

Bernheimer, H. (1964). Distribution of homovanillic acid in the human brain. Nature 204, 587-588.

Bertler,A. (1961). Occurrence and localization of catecholamines in the human brain. Acta physiol.scand. 51, 97-107.

Birkmayer, W., Danielczyk, W., Neumayer, E., Riederer, P. (1974). Nucleus ruber and L-dopa psychosis: biochemical post-mortem findings. J.Neural Transm. 35, 93-116.

Birkmayer, W., Riederer, P., Youdim M.B.H., Linauer W. (1976). Potentiation of anti akinetic effect after L-dopa-treatment by an inhibitor of MAO-B, Deprenil. In Advances in Parkinsonism (eds.W.Birkmayer, O.Hornykiewicz), pp 381-396, Editiones Roche, Basle.

Commissiong, J.W., Galli, C.L., Neff, N.H. (1978). Differentiation of dopaminergic and noradrenergic neurons in rat spinal cord. J.Neurochem. 30, 1095-1099.

Commissiong, J.W., Neff, N.H. (1979). Current status of dopamine in the mammalian spinal cord. Biochem.Pharmacol. Vol.28, 1569-1573.

Dahlström, A., Fuxe, K. (1964). Evidence for the existence of monoamine-containing neurons in the central nervous system. I. Demonstration of monoamines in the cell bodies of the brainstem neurons. Acta physiol.scand. 62, 1-55.

DeArmond, St.J., Fusco, M.M., Dewey, M.M. (1976). Structure of the human brain. A photographic atlas. 2nd ed. New York-Oxford, University Press

Ehringer H., Hornykiewicz, O. (1960). Verteilung von Noradrenalin im Dopamin (3-hydroxytryptamin) im Gehirn des Menschen und ihr Verhalten bei Erkrankungen des extrapyramidalen Systems. Klin.Wschr. 38, 1236-1240.

Hjemdahl, P., Daleskog, M., Kahan, T. (1979). Determination of plasma catecholamines by high performance liquid chromatography with electrochemical detection. Life Science 25, 131-138.

Hornykiewicz, O., Lisch, H.J., Springer, A. (1968). Homovanillic acid in different regions of the human brain: Attempt at localizing central dopamine fibres. Brain Res. 11, 662-671.

Jellinger, K., Riederer, P. (1977). Brain monoamines in metabolic (endotoxic) coma. A preliminary biochemical study in human post-mortem material. J.Neural Transm. 41, 275-286.

Keller, R., Oke, A., Mefford, I., Adams, R.N. (1976). Liquid chromatographic analysis of catecholamines. Routine assay for regional brain mapping. Life Sci. 19, 995-1004.

Massion J. (1967). The mammalian red nucleus. Physiological Rev. 47, 383-436.

Miller, R.A., Strominger, N.L. (1973). Efferent connections of the red nucleus in the brainstem and spinal cord of the rhesus monkey. J.Comp.Neur. 152, 327-346.

Moses, S.G., Robins, E. (1975). Regional distribution of norepinephrine and dopamine in brains of depressive suicides and alcoholic suicides. Psychopharmacol. Commun. 1, 327-337.

Nieuwenhuys, R., Voogd, J., van Huijzen, Chr. (1979). The human central nervous system. A synopsis and atlas. Springer, Berlin-Heidelberg-New York.

Ovtscharoff, W. (1972 a). Histochemie und Elektronenmikroskopie des Nucleus ruber der Schildkröte (Testudo Hermanni) Histochemie 29, 240-247.

Ovtscharoff, W. (1972 b). Zur Histogenese und Chemodifferentierung des Nucleus ruber der Ratte. Histochemie 29, 220 - 239.

Poitras, D., Parent, A. (1978). Atlas of the distribution of monoamine-containing nerve cell bodies in the brain stem of the cat. J.Comp.Neurol. 179, 699-717.

Riederer, P., Wuketich, St. (1976). Time course of nigrostriatal degeneration in Parkinson's disease. J.Neural Transm. 38, 277-301.

Riederer, P., Rausch, W.-D., W.Birkmayer, K.Jellinger, D.Seemann (1978). CNS Modulation of adrenal tyrosine hydroxylase in Parkinson's disease and metabolic encephalopathies. J.Neural Transm., Suppl.14, 121-131.

Sano, I., Gamo, T., Kakimoto, Y., Takesada, M., Nishinuma,K. (1959). Distribution of catechol compounds in human brain. Biochim.Biophys. Acta 32, 586-587.

Shimada, S., Ishikawa, M., Tanaka, C. (1976). Histochemical mapping of dopamine neurons and fiber pathways in dog mesencephalon. J.Comp.Neurol. 168, 533-543.

Wuketich, St., Riederer, P., Jellinger, K., Ambrozi, L. (1980). Quantitative Dissection of human brain areas: Relevance to transmitter analyses. J.Neural Transm., Suppl.16, 53-67.

125

Function and Dysfunction of the GABA System in the Human Brain

K. G. LLOYD*, C. MUNARI†, L. BOSSI*, J. BANCAUD†,
J. TALAIRACH and P. L. MORSELLI*

* Research Department, Synthélabo—L.E.R.S.,
58 Rue de la Glacière, Paris 75013, France
† Unité 97 INSERM, Hôpital Ste Anne, Paris, France

INTRODUCTION

It has been postulated for some time that GABA (gamma-aminobuty-
ric acid) functions as the major inhibitory neurotransmitter in
the brain (cf. Roberts, 1976). Over the past 5-6 years the interest in GABAergic function has increased enormously, with at
least 7 symposia or books devoted specifically to this neurotransmitter (Roberts et al, 1976 ; Fonnum, 1978 ; Krogsgaard-Larsen et
al, 1979 ; Mandel and De Feudis, 1979 ; 1980 ; Fielding and Lal,
1980 ; Costa et al, 1980) not to mention other speciality symposia where GABA has received major attention (eg. see Chase et al,
1979 ; Poirier et al, 1979 ; Yamamura et al, 1980). The amount of
information available on central GABA systems is almost reaching
overload proportions in all fields (eg. neurophysiology, neurochemistry, neuropharmacology, neuroendocrinology). The result of
this is that a picture is beginning to emerge as to which human
CNS disorders may involve a GABA neuron dysfunction and therefore may be amenable to treatment by GABAergic drugs. From these
various studies and from the sparse clinical information available (Shoulson et al, 1975 ; Bartholini et al, 1979a ; Chase and
Tamminga, 1979 ; Morselli et al, 1980) it appears that abnormal
GABAergic function is involved in the genesis of several neurological diseases, but that convincing evidence for GABAergic dysfunction in psychiatric states is still lacking. The present
review will concentrate on those disorders in which sufficient
evidence from human brain material exists to reasonably propose
a dysfunction of GABA neurons. These states are : Parkinson's
disease, Huntington's disease and epilepsy.

METHODS AND MATERIALS

Human brain material was obtained, frozen and dissected as previously described (Lloyd et al, 1975, 1980a), GAD (glutamic acid decarboxylase) activity was performed as described by Lloyd and Hornykiewicz (1973) for Parkinson's disease material and by Bird and Iversen (1974) for the epilepsy material. ^{3}H-GABA binding was analyzed according to Lloyd et al (1977a) and GABA-T (GABA-transaminase) by the method of Wu (1976).

INDICES OF GABA SYNAPTIC FUNCTION IN BRAINS OF "NORMAL" PATIENTS

In the brains of patients who died without apparent neurological or psychiatric disease, the biochemical indications for the presence of GABAergic neurons are similar to those seen in other mammalian species. Thus, GAD, the enzyme responsible for the synthesis of GABA, the GABA receptor (as determined by ^{3}H-GABA binding) and GABA levels are all present in different human brain regions (Table 1). The results presented in Table 1 are similar to those reported from other laboratories ; for GAD (Bird and Iversen, 1974 ; McGeer et al, 1971 ; Rinne et al, 1974 ; cf. Lloyd, 1972 for earlier references) ; for ^{3}H-GABA binding (Enna et al, 1977 ; Rinne et al, 1978 ; Olsen et al, 1980) ; and for GABA levels (Perry et al, 1971).

TABLE 1. Indices of GABA synaptic activity in "normal" human brain

Brain Region	GAD Activity[a] nmol CO_2/100 mg protein/hr	GABA Levels[a] µg/g Tissue	H-GABA Binding[b] fmol/mg Protein (no Triton-X-100)
Caudate Nucleus	1318 ± 186 (13)	179 (2)	66 ± 8 (9)
Putamen	1243 ± 220 (13)	191 (2)	68 ± 14 (17)
Pallidum (Ext)	1106 ± 306 (8)	431 (1)	16 ± 3 (6)
Substantia Nigra	1273 ± 297 (13)	–	31 ± 5 (11)
Temporal Cortex	1024 ± 154 (13)	112 (1)	188 ± 28 (12)
Cerebellar Cortex	780 ± 149 (12)	100 (1)	328 ± 50 (10)

Data expressed as mean (with SEM if more than 3 brains). Number of brains examined in parentheses. a. Data from Lloyd, 1972 ; b. Data from Lloyd and Dreksler, 1978.

Not only are these parameters present, but within the limits of the experiments performed, the pharmacology of the human GABA synapses are similar to those observed in the brains of laboratory animals. Thus, GAD activity is inhibited by cycloserine "ex vivo" both in the human (Table 2) and in the rat brain (Dengler et al, 1962). In this case, it is worth noting that although GAD activity is reduced by 80 percent throughout the brain, GABA levels (at least in the pallidum and cerebellum) are normal, indicating a large functional excess of GAD activity.

TABLE 2. Inhibition of GAD in the brain of a patient who received Cycloserine

| Region | Glutamic Acid Decarboxylase Activity | | |
| | Control Values | Cycloserine Patient | |
	nmol CO_2 hr^{-1} 100 mg Protein	nmol CO_2 x hr^{-1} 100 mg Protein	Percent Control
Caudate N.	1318 ± 186 (13)	228	17
Putamen	1243 ± 220 (13)	164	13
Pallidum (Ext)	1106 ± 306 (8)	268	18
Hypothalamus	1241 ± 141 (4)	233	19
Temporal Cortex	1024 ± 154 (13)	200	20
Cerebellar Cortex	780 ± 149 (12)	140	18
Substantia N.	1273 ± 297 (13)	226	18

Results of control patients expressed as mean with SEM. Number of patients in parentheses.

The pharmacology of the GABA receptor in the human brain (as defined by ^{3}H-GABA binding) has been well delineated. Thus, the affinity of the receptor for GABA is greatly enhanced by Triton-X-100 treatment (Enna et al, 1979 ; Lloyd and Dreksler, 1979) ; ^{3}H-GABA is specifically displaced by drugs which are active neurophysiologically at the GABA receptor, eg muscimol, 3-aminopropanesulfonic acid, imidazoleacetic acid, SL 75 102 and bicuculline (Bartholini et al, 1979a ; Enna et al, 1979 ; Lloyd and Dreksler, 1978 ; 1979 ; Iversen, 1978). Compounds such as GABA uptake inhibitors, GABA-transaminase inhibitors, baclofen, picrotoxinin and the barbiturates are ineffective on ^{3}H-GABA binding to human cerebellar membranes (Enna et al, 1979 ; Lloyd and Dreksler, 1978 and 1979 ; Iversen, 1978).

GABA NEURONAL FUNCTION IN PATIENTS WITH PARKINSON'S DISEASE

The major neurochemical dysfunction in Parkinson's disease is the well-known loss of dopaminergic neurones, notably in the nigrostriatal dopamine pathway (cf. Hornykiewicz, 1966 ; Lloyd et al, 1975). However, the chronic Parkinsonian condition is more complex than a simple loss of DA neurons. As far as the GABA system is concerned, all studies to date have noted at least a 50 percent loss of GAD activity in the extrapyramidal system of non-DOPA treated Parkinsonian patients (Table 3) ; GABA levels themselves are not altered (Table 3). This latter observation may be a result of post-mortem artefacts or may reflect the possibility that the GABA neurons have not actually degenerated but are rather in a metabolically depressed state following a prolonged reduction in dopamine input. This latter hypothesis is supported by the observations that (i) lesions of nigro-striatal dopamine neurons result in a large decrease in the metabolic activity of the striatum (Schwartz et al, 1976) ; and (ii) GAD levels of chronically

L-DOPA treated patients are relatively "normal" (Table 3). This has been interpreted as indicative of a reversal of a "biochemical atrophy" of GABAergic neurons in the untreated Parkinsonian condition, as if GABAergic neurons had actually been lost, L-DOPA therapy would not be associated with normalized levels of GAD (cf. Lloyd, 1980).

TABLE 3. GAD activity, GABA levels and ^{3}H-GABA binding in different brain regions from Parkinsonian patients

| Brain Region | GAD Levels[a] | | GABA Levels | ^{3}H-GABA Binding |
	No DOPA	L-DOPA Treated	No DOPA	All Patients
Caudate Nucleus	$25^b; 31^c$	$73^c; 78^b; 101^d$	75^e	$94^h; 107^g$
Putamen	$30^b; 50^c$	$75^c; 78^b; 108^d$	104^e	$95^h; 135^g$
Substantia Nigra	$20^b; 52^c$	$26^d; 41^c; 95^b$	137^f	$31^g; 54^h$

a. All values expressed as percent of results from control patients. Data from : b. Lloyd et al, 1976 ; c. McGeer et al, 1971 (all patients) ; d. Rinne et al, 1974 ; e. Hornykiewicz et al, 1976 ; f. Perry et al, 1973 ; g. Lloyd et al, 1977a ; h. Rinne et al, 1978.

The question of the clinical correlates of these alterations in GAD activity in the extrapyramidal system is of course of interest. It has been suggested that the decrease in GAD activity in the untreated Parkinsonian condition is a homeostatic manoeuver to maintain the GABA/dopamine balance in the extrapyramidal system and that upon reinstatement of the dopaminergic tone by chronic L-DOPA therapy the GABA function also returns to normal (cf. Lloyd, 1980). However other interpretations or correlations are also possible. Thus, the tremor of Parkinson's disease apparently is not correlated with the dysfunction of dopamine neurons (Bernheimer et al, 1973), and responds only poorly to acute L-DOPA therapy (Barbeau, 1969). Upon prolonged treatment with L-DOPA, the Parkinsonian tremor undergoes a marked amelioration (Barbeau, 1969 ; Calne and Klawans, 1977), a time-course which has an evident resemblance to that for L-DOPA-GAD effect. Additional indirect evidence is available for the involvement of striatal GABA neurons in tremor. Thus, one of the most efficacious treatments for Parkinsonian tremor is the administration of anticholinergic drugs (Calne and Reid, 1972 ; Greenblatt and Shader, 1973). Increased GABAergic tone (eg by progabide or muscimol) effectively diminished striatal cholinergic activity (Scatton and Bartholini, 1980). This may be correlated to the L-DOPA-GAD-antitremor time course of action.

The above discussion applies to the pathology of GABA neurons, i.e. the presynaptic element. Changes in GABA receptors in the extrapyramidal system have also been noted in the Parkinsonian condition. In the substantia nigra a loss of 50-70 percent of ^{3}H-GABA binding sites is evident (Table 3). This is likely at least partially associated with the loss of the dopamine cell bodies in this region as, firstly, this latter change is the quantitatively most important morphological and neurochemical alteration to occur in this nucleus (cf. Hornykiewicz, 1966 ; Lloyd et al, 1975), and secondly, such a loss in ^{3}H-GABA binding does not occur in the striatum (Table 3) where decreases in serotonin occur similar to those observed in the substantia nigra (cf. Hornykiewicz, 1966).

The evidence available from animal experiments supports the hypothesis that GABA neurons exhibit a tonic inhibitory control on dopamine neuron function both at the level of the substantia nigra and the corpus striatum (cf. Lloyd et al, 1979a ; Worms et al, 1979). As (i) in Parkinson's disease GAD activity is reduced throughout the extrapyramidal system and (ii) the density of ^{3}H-GABA binding sites is decreased in the substantia nigra, it can be proposed that the GABA-mediated tonic inhibition of nigro-striatal dopamine neurones is reduced. There is good evidence that in Parkinson's disease (Lloyd and Davidson, 1979a) or 6-hydroxy-dopamine lesioned rats (Agid et al, 1973) the remaining nigro-striatal dopamine neurons are maximally activated. It has been proposed that one of the mechanisms of activation of the nigro-striatal neurons is a disinhibition via the striato-nigral GABA input and perhaps also the GABA-dopamine inhibition at the striatal level (Lloyd and Davidson, 1979a ; Lloyd and Hornykiewicz, 1973). Both post-mortem findings (cf. Lloyd, 1977a ; Lloyd, 1980) and animal models (cf. Lloyd et al, 1979· , ; Kaäriainen, 1976 ; Delini-Stula, 1979) support this hypothesis. This may predict that non-convulsant GABA antagonists may be efficacious as anti-Parkinsonian agents.

The emergence of L-DOPA-induced dyskinesias as the limiting factor in the long-term L-DOPA treatment of Parkinson's disease may involve further alterations in the GABA/dopamine balance and thus be amenable to therapy by GABA-related drugs. In an animal model of L-DOPA dyskinesias (apomorphine stereotypies) progabide, a direct-acting GABA mimetic, exerts a marked antidyskinetic effect (Bartholini et al, 1979b). Preliminary clinical observations show that progabide eliminates L-DOPA dyskinesias and that excess doses of this GABA-mimetic drug will aggrevate the underlying Parkinsonism (Bartholini et al, 1979a ; Morselli et al, 1980).

HUNTINGTON'S DISEASE

Huntington's disease has been frequently considered to be the clinical "mirror image" of Parkinson's disease. Thus, Huntington's

131

disease is characterized by massive cortical and striatal degeneration, hyperkinesias and a prevalence of psychosis and dementia (cf. Barbeau et al, 1973) whereas Parkinson's disease exhibits akinesia, rigidity and, if a psychiatric correlate exists, depression (cf. Hornykiewicz, 1966 ; Lloyd, 1977b). In Parkinson's disease little striatal cell loss is consistently observed (Oppenheimer, 1976). Pharmacotherapy reflects this difference in symptoms as neuroleptics are efficacious in Huntington's disease but induce Parkinson's disease (cf. Ringel et al, 1973 ; Chase, 1976), whereas L-DOPA is beneficial in Parkinson's disease but detrimental in Huntington's disease (cf. Calne and Reid, 1972 ; Fahn et al, 1973).

TABLE 4. Indices of GABA synaptic function in post mortem basal ganglia material from Huntington's chorea patients

Index of GABA Synaptic Function Percent Control Patients	Caudate Nucleus	Putamen	Substantia Nigra
GAD Activity	$19^a; 27^b; 38^c$	$10^a; 19^b; 32^c$	31^b
GABA Levels	$43^c; 60^d$	$38^e; 41^c$	40^d

Data from ; a. Stahl and Swanson (1974) ; b. McGeer and McGeer (1976) ; c. Bird and Iversen (1974) ; d. Perry et al (1973).

GABA neurons in Huntington's disease undergo a severe degeneration as not only are GAD concentrations considerably diminished, but GABA levels themselves are very low in the extrapyramidal system (Table 4). The functional state of GABA receptors (as defined by ^{3}H-ligand binding) in Huntington's disease (Table 5) has been the subject of some debate. Although initial studies indicated that ^{3}H-GABA binding is normal in the caudate nucleus and putamen (Enna et al, 1976a, b), this finding was soon challenged (Lloyd et al, 1977b). It has now been confirmed that ^{3}H-GABA (or ^{3}H-muscimol) binding is significantly decreased in striatal regions from Huntington's disease patients (Table 5). This finding is consistent with the massive cell loss and gliosis observed post-mortem in these regions in Huntington's disease brains (Barbeau et al, 1973). This finding has clinical significance in that it implies that the success of GABA agonists in some patients with Huntington's disease (Bartholini et al, 1979a; Morselli et al, 1980) is due either to extra-striatal actions of these drugs, or that the remaining GABA receptors are supersensitive. The question of extrastriatal GABA receptors and striatal function is important, but unresolved. The only immediately relevant information appears to be (i) an efferent pathway from the substantia nigra controlled, inter alia, by the striatonigral tract (DiChiara et al, 1979 ; Scheel-Kruger, 1980) and (ii) GABA receptors regulate DA-mediated functions post-synaptic to the striatal DA receptor (Bartholini, 1980 ; Lloyd and Worms, 1979), but whether this occurs extra-or-intrastriatally is unknown.

TABLE 5. ^{3}H-GABA and ^{3}H-muscimol binding to membranes prepared from different brain regions of patients with Huntington's disease

^{3}H-Ligand Used	Caudate Nucleus	Putamen	Substantia Nigra	Frontal Cortex	Cerebellar Cortex
GABA	73[a] 22[b]** 15[c]**	109[e] 50[f]* 25[b]**	183[e] 115[c]	105[a] 103[c]	194[b] 91[c] 115[f]
Muscimol	64[d]**	67[d]**		107[d]	

Values are expressed as percent of control patients. ^{3}H-ligand binding performed at one concentration (4-25 nM) under different preparation procedures. * $p < 0.05$; ** $p < 0.01$ vs control. a. Enna et al (1976a) ; b. Lloyd et al, 1977b ; c. Olsen et al, 1980 ; d. Reisine et al, 1979 ; e. Enna et al (1976b) ; f. Iversen et al, 1979.

The other possibility, that the remaining GABA receptors become supersensitive is quite likely. Thus, the cerebellar cortex from Huntington's patients is used as a model for striatal changes as (1) there is usually insufficient striatal tissue present for kinetic studies and (2) qualitatively similar neurochemical changes are observed in both structures (cf. Lloyd and Davidson, 1979a and b). In membranes prepared from cerebellar cortex the apparent Kd for ^{3}H-GABA binding is 3-fold lower in Huntington's disease than in control patients (Table 6). This alteration in Kd is not limited only to the cerebellar cortex but also occurs in cerebral cortical regions (Lloyd et al, 1980b). If a similar kinetic alteration occurs in the striatum this enhanced affinity may increase the efficacy of GABA-mediated transmission towards a normal function. It should be noted that the apparently normal ^{3}H-GABA binding shown for the frontal cortex in Table 5 is due to the combination of an increased affinity and decreased Bmax.

Further studies on these GABA receptors indicate that this kinetic change may be due to an alteration in membrane properties, possibly a phospholipid. Thus, Triton-X-100 or phospholipase-C (both of which normally decrease the Kd for ^{3}H-GABA by 3-4 fold ; Table 6) do not further alter the ^{3}H-GABA binding (Kd) to cerebellar membranes from Huntington's disease patients (Table 6). This may indicate that the change induced by these substances has already occurred in Huntington's diseases brains.

TABLE 6. The effect of preincubation of Triton-C-100 or Phospho-
lipase C on the ^{3}H-GABA binding to membranes prepared
from control or Huntington's patients

Incubation at 37° for 30 min	Control Patients (n = 8)		Huntington's Patients (n = 8)	
	Kd (nM)	Bmax (fmol/mg protein)	Kd (nM)	Bmax (fmol/mg protein)
Untreated Membranes	101 ± 8	76 ± 13	37 ± 4**	90 ± 8
Triton-X-100 (0.02%)	38 ± 3	98 ± 15	31 ± 4	131 ± 15
Phospholipase C (1 m Unit)	44 ± 11++	67 ± 9	31 ± 4	107 ± 11

After preincubation membranes were washed 3 times and then
resuspended in buffer for the ^{3}H-GABA binding assay. Results
expressed as mean with SEM. Data from Lloyd and Davidson, 1979b.
**$p < 0.01$ vs corresponding material from Control Patients.
++ $p < 0.01$ vs corresponding untreated material.

EPILEPSY

The observations that most antiepileptic drugs augment GABA
synaptic activity and that most GABA mimetic drugs are anti-
convulsant, together with the convulsant effects obtained by
a blockade of GABA synaptic activity (for reference cf. Lloyd et
al, 1980a), have led to the hypothesis that (i) a dysfunction of
GABA neurons may be causative to certain epileptic states and
(ii) that non-toxic GABA agonists will be effective antiepileptic
drugs. While the second question is outside the limits of this
discussion, it can be noted that progabide, a GABA agonist, has
been effective as an anti-epileptic drugs in clinical trials to
date (Bartholini et al, 1979a ; Lloyd et al, 1980a ; Morselli
et al, 1980).

A direct verification of the GABA hypothesis of epilepsy would be
to examine material removed from the brains of epileptic patients
either at the time of surgical resection for uncontrolled epilepsy or post-mortem. Some initial studies have been performed
(cf. Tower, 1976) but the results have been equivocal. We have
presently reexamined the question of GABA neuron dysfunction
using material obtained at surgical resection from epileptic
patients with previous stereo-EEG analysis by means of deep-
implanted electrodes. Furthermore, for many samples a neuroanatomical analysis of the tissue has been made.

The major findings to date from this study are shown in Table 7.
GAD activity was found to be decreased in epileptic regions as
compared to non-epileptic regions of the same brains ; this is
true whether the reference criterion was a neurophysiological
parameter (stereo-EEG recordings from depth electrodes) or neuroanatomical parameters. When the GAD activities were analyzed from

134

individual patients (data not shown), several further conclusions
can be drawn : (1) all patients with epilepsy due to the presence
of a tumour show a marked reduction in GAD activity in the affected areas ; (2) for patients with epilepsy not associated with
the presence of a tumour, GAD activity in epileptic regions varied
greatly between patients. In those patients with a reduced GAD
activity (about half of the patient group) this was true for all
epileptic regions. This indicates that some epileptic states are
associated with a decreased GABA neuron activity.

TABLE 7. Neurochemical alterations observed in samples from
surgically resected temporal cortex from epileptic
patients[1]

	Glutamic Acid Decarboxylase nmol CO_2/mg protein/hr	^{3}H-GABA Binding		GABA Transaminase nmol/succinate/ mg protein/15min
		Kd nM	^{3}H-GABA Bound fmol/mg Protein	
Regions without EEG or Neuro-anatomical Abnormalities	79 ± 9 (23)	89±5 (7)	139 ± 20 (11)	2.0 ± 0.2 (22)
Epileptogenic[2] Brain Regions				
Tumour Patients	33 ± 8*** (5)	–	–	2.6 ± 0.5 (5)
Non-Tumour Patients	53 ± 4** (18)	101±10 (7)	104 ± 13 (11)	2.4 ± 0.2 (17)
Anatomo-pathological Brain Regions[3]				
Tumour Patients	30 ± 5*** (5)	–		2.6 ± 0.4 (5)
Non-Tumour Patients	57 ± 5* (18)	–		2.4 ± 0.7 (17)

1. Results expressed as mean with S.E.M. Number of patients in
parentheses. 2. As defined from depth electrodes. 3. As defined
from gross microscopy studies (scarring, malformation, gliosis,
etc). * $p < 0.05$; ** $p < 0.01$; *** $p < 0.001$ vs control regions

The binding of ^{3}H-GABA was studied with respect to two parameters : the affinity for the binding site (Kd) and the amount
bound at a single concentration (25 nM) of ^{3}H-GABA. At present
only the results for stereo-EEG analysis of non-tumour patients
are available for comparative analysis. When the data from all
of the patients are globally analyzed (Table 7) the Kd or amount
bound in the epileptic regions are not different from that ob-

served for the non-epileptic areas. However when the cases are analyzed individually it is apparent that there is a marked patient-to-patient variation in both parameters. Together with the large inter-individual variations in the loss of GAD activity in the epileptic zones, this may explain why different blood concentrations (and therefore probably brain levels) of progabide are needed to obtain an antiepileptic action (Morselli et al, 1980). If this is correct, patients with fewer functional GABA terminals and/or receptors would need higher doses of GABA agonists than those with more normal GABA synapses. GABA-transaminase activity was not significantly altered in these brain regions (Table 7). This was the case when either tumour or non-tumour patients were examined ; when the patients were individually examined only one of 22 cases showed a low GABA-T activity in epileptic regions. This finding tends to confirm the hypothesis that the alterations in GAD activity and GABA binding are specific for GABA synapses and are not generally a result of a non-specific tissue loss.

SUMMARY AND CONCLUSIONS

From the data presently reviewed, it can be concluded that extra-pyramidal system GABA neuron dysfunction is likely associated with at least three major neurological disorders : Parkinson's disease, Huntington's disease and epilepsy. However, the underlying basis for the decreases in GABA synaptic activity is not the same in all cases. Thus, Huntington's disease and some cases of epilepsy appear to be associated with a primary deficit in GABA synaptic function whereas in Parkinson's disease the decrease in striatal GABA neuron activity may be a homeostatic compensation to the primary dopamine neuron loss. From this one would suggest that a non-toxic GABA agonist would be beneficial in Huntington's disease and epilepsy but may exacerbate Parkinsonism. In the latter case, a non-convulsant GABA antagonist (if one can be found) may be of use.

This list of neurological diseases is by no means exhaustive, and there is preliminary evidence that in other disorders such as alzheimer's dementia (Reisine et al, 1978) GABAergic function may be abnormal. For psychiatric disorders there is as yet no firm biochemical data indicating a GABA neuron dysfunction (Crow et al, 1978 ; Iversen et al, 1979). However, GABA neuron function may be involved in some of the effects of drugs used in psychiatric disorders (cf. Lloyd, 1980 ; Morselli et al, 1980).

REFERENCES

Agid, Y., Javoy, F. and Glowinski, J. (1973). Hyperactivity of Remaining Dopaminergic Neurons After Partial Destruction of the Nigrostriatal Dopaminergic System in the Rat. Nature New Biol. 245, 150-151.

Barbeau, 1. (1969) L-DOPA Therapy in Parkinson's Disease : A Critical Review of Nine Years' Experience. Canad. Med. Assoc. J., 101, 59-68.

Barbeau, A., Chase, T. N. and Paulson, G. W. (1973). Huntington's Chorea : 1872-1972. Advances in Neurology, vol. 1, Raven Press, New York.

Bartholini, G. (1980) Interaction of Striatal Dopaminergic, Cholinergic and GABA-ergic Neurons : Relation to Extrapyramidal Function. Trends Pharmacol. Sci., 1, 138-141

Bartholini, G. B., Scatton, B., Zivkovic, B. and Lloyd, K. G. (1979a). On the Mode of Action of SL 76 002, A new GABA Receptor Agonist. In GABA-Neurotransmitters, (eds P. Krogsgaard-Larsen, J. Scheel-Kruger and H. Kofod), Munksgaard, Copenhagen.

Bartholini, G., Lloyd, K. G., Worms, P., Constantinidis, J. and Tissot, R. (1979b) GABA and GABA-ergic Medication : Relation to Striatal Dopamine Function and Parkinsonism : In Parkinson's Disease, Adv. in Neurology, vol. 24 (eds L. J. Poirier, T. L. Sourkes and P. J. Bédard) Raven Press, New York.

Bernheimer, H., Birkmayer, W., Hornykiewicz, O., Jellinger, K. and Seitelberger, F. (1973) Brain Dopamine and the Syndromes of Parkinson and Huntington. J. Neurol. Sci., 20, 415-455.

Bird, E. D. and Iversen, L. L. (1974). Huntington's Chorea : Post Mortem Measurement of Glutamic Acid Decarboxylase, Choline Acetylase and Dopamine in Basal Ganglia. Brain, 97, 457-472.

Calne, D. B. and Klawans, H. L. (1977). Pathophysiology and Pharmacotherapy of Tremor. Pharmacol. Therap. C. 2, 113-123.

Calne, D. B. and Reid, J. L. (1972). Antiparkinsonian Drugs : Pharmacolog·cal and Therapeutic Aspects. Drugs,4, 49-74.

Chase, T. N. (1976). Antipsychotic Drugs, Dopaminergic Mechanisms and Extrapyramidal Function in Man. In Antipsychotic Drugs ; Pharmacodynamics and Pharmacokinetics (eds G. Sedvall, B. Uvnas and Y. Zotterman), Pergamon Press, New York.

Chase, T. N. and Tamminga, C. A. (1979). GABA System Participation in Human Motor, Cognitive and Endocrine Function. In GABA-Neurotransmitters (eds P. Krogsgaard-Larsen, J. Scheel-Kruger and H. Kofod), Munksgaard, Copenhagen.

Chase, T. N., Wexler, N. S. and Barbeau, A. (1979). Huntington's Disease, Raven Press, New York.

Costa, E., DiChiara, G. and Gessa, G. L. (1980) GABA and Benzodiazepine Receptors, Raven Press, New York.

Crow, T. J., Owen, F., Cross, A. J., Lofthouse, R. and Longden, A. (1978). Brain Biochemistry in Schizophrenia. Lancet, i, 36-37.

Delini-Stula, A. (1979) Differential Effects of Baclofen and Muscimol on Behavioural Responses Implicating GABA-ergic Transmission : In GABA-Neurotransmitters (eds P. Krogsgaard-Larsen, J. Scheel-Kruger and H. Kofod) Munskgaard, Copenhagen.

Dengler, H. J., Rauchs, E. and Rummel, W. (1962). Zur Hemmung der L-Glutaminsaure- und L-Dopadecarboxylase durch D-Cycloserin und andere Isoxazolidone. Arch. Exp. Path., 243, 366-381.

137

DiChiara, G., Porceddu, M. L., Morselli, M., Mulas, M. L. and Gessa, G. L. (1979). Strio-nigral and Nigro-thalamic GABAergic Neurons as Output Pathways for Striatal Responses : in GABA-Neurotransmitters (Eds P. Krogsgaard-Larsen, J. Scheel-Kruger and H. Kofod), Munksgaard, Copenhagen.

Enna, S. J., Bird, E. D., Bennett, J. P., Bylund, D. B., Yamamura, H. I., Iversen, L. L. and Snyder, S. H. (1976a). Huntington's Chorea : Changes in Neurotransmitter Receptors in the Brain. New Engl. J. Med., 294, 1305-1309.

Enna, S. J., Bennett, J. P., Bylund, D. B., Snyder, S. H., Bird, E. D. and Iversen, L. L. (1976b). Alterations of Brain Neurotransmitter Receptor Binding in Huntington's Chorea. Brain Res., 116, 531-537.

Enna, S. J., Bennett, J. P., Bylund, D. B., Creese, I., Burt, D. R., Charness, M. E., Yamamura, H. I., Simantov, R. and Snyder, S. H. (1977) Neurotransmitter Receptor Binding : Regional Distribution in Human Brain. J. Neurochem., 28, 233-236.

Enna, S. J., Ferkany, J. W. and Krogsgaard-Larsen, P. (1979). Pharmacological Characteristics of GABA Receptors in Different Brain Regions. In GABA-Neurotransmitters (eds P. Krogsgaard-Larsen, J. Scheel-Kruger and H. Kofod), Munksgaard, Copenhagen.

Fahn, S., Mishkin, M. M. and Hoffman, R. R. (1973).Pharma-cologic and Radiologic Investigations in Huntington's Chorea : In Huntington's Chorea : 1872-1972. Advances in Neurology, vol. 1 (eds A. Barbeau, T. N. Chase and G. W. Paulson), Raven Press, New York.

Fielding, S. H. and Lal, H. (1980) GABA and Other Inhibitory Transmitters, ANKHO, Fayetteville.

Fonnum, F. (1978) Amino Acids as Chemical Transmitters, Plenum Press, New York.

Greenblatt, D. J. and Shader, R. I. (1973) Anticholinergics. New Engl. J. Med., 288, 1215-1219.

Hornykiewicz, O. (1966) Dopamine (3-hydroxytyramine) and Brain Function. Pharmacol. Revs., 18, 925-964.

Hornykiewicz, O., Lloyd, K. G. and Davidson, L. (1976). The GABA System Function of the Basal Ganglia and Parkinson's Disease. In GABA in Nervous System Function (eds E. Roberts, T. N. Chase and D. B. Tower), Raven Press, New York.

Iversen, L. L. (1978). Biochemical Psychopharmacology of GABA. In Psychopharmacology : A Generation of Progress (eds M. A. Lipton, A. DiMascio and K. F. Killam), Raven Press, New York.

Iversen, L. L., Bird, E. D., Spokes, E., Nicholson, S. H. and Suckling, C. J. (1979). Agonist Specificity of GABA Binding Sites in Human Brain and GABA in Huntington's Disease and Schizophrenia. In GABA-Neurotransmitters (eds P. Krogsgaard-Larsen, J. Scheel-Kruger and H. Kofod) Munksgaard, Copenhagen.

Kääriäinen, I. (1976). Effects of Aminooxyacetic Acid and Baclofen on the Catalepsy and on the Increase of Mesolimbic and Striatal Dopamine Turnover Induced by Haloperidol in Rats. Acta Pharmacol. Toxicol., 39, 393-400.

Krogsgaard-Larsen, P., Scheel-Kruger, J. and Kofod, H. (1979) GABA-Neurotransmitters, Munksgaard, Copenhagen.

Lloyd, K. G. (1972) Biogenic Amines and Related Enzymes in the Human and Animal Brain. Ph. D. Thesis, University of Toronto.

Lloyd, K. G. (1977a). Neurochemical Compensation in Parkinson's Disease. In Parkinson's Disease : Concepts and Prospects (eds J.P.W.L. Lakke, J. Korf and H. Wesseling), Excerpta Medica, Amsterdam.

Lloyd, K. G. (1977b). Psychiatric Disturbances Occurring During Levodopa Therapy of Parkinson's Disease. Prim. Care, 4, 561-575.

Lloyd, K. G. (1980) . Indications for GABA Neuron Dysfunction in Mental Disease. In Enzymes and Neurotransmitters in Mental Disease (eds. E. Ussin, T.L. Sourkes and M.B.H. Youdim), John Wiley, London.

Lloyd, K. G. and Davidson, L. (1979a) Involvement of GABA Neurons and Receptors in Parkinson's Disease and Huntington's Chorea : A Compensatory Mechanism ? In Parkinson's Disease : Advances in Neurology, Vol. 24 (eds L. J. Poirier, T. L. Sourkes and P. J. Bédard) Raven Press, New York.

Lloyd, K. G. and Davidson, L. (1979b). ^{3}H-GABA Binding in Brains from Huntington's Chorea Patients : Altered Regulation by Phospholipids . Science, 205, 1147-1149.

Lloyd, K. G. and Dreksler, S. (1978). ^{3}H-GABA Binding to Membranes Prepared from Post-Mortem Human Brain : Pharmacological and Pathological Investigations. In Amino Acids as Chemical Transmitters (ed. F. Fonnum) Plenum Press, New York.

Lloyd, K. G. and Dreksler, S. (1979). An Analysis of ^{3}H-Gamma Aminobutyric Acid (GABA) Binding in the Human Brain. Brain Res., 163, 77-87.

Lloyd, K. G. and Hornykiewicz, O. (1973). L-Glutamic Acid Decarboxylase in Parkinson's Disease : Effect of L-DOPA Therapy. Nature, 243, 521-523.

Lloyd, K. G. and Worms, P. (1979). Sustained GABA Receptor Stimulation and Chronic Neuroleptic Effects. In Long Term Effects of Neuroleptics (eds F. Cattabeni, G. Racagni and P. F. Spano), Raven Press, New York.

Lloyd, K. G., Davidson, L. and Hornykiewicz, O. (1975). The Neurochemistry of Parkinson's Disease : Effect of L-DOPA Therapy, J. Pharmacol. Exp. Therap. 195, 453-464.

Lloyd, G. L., Möhler, H., Bartholini, G. and Hornykiewicz, O. (1976). Pathological Alterations in Glutamic Acid Decarboxylase Activity in Parkinson's Disease. In Advances in Parkinsonism (eds W. Birkmayer and O. Hornykiewicz) Editions Roche, Basel.

Lloyd, K. G., Shemen, L. and Hornykiewicz, O. (1977a). Distribution of High Affinity Sodium-Independent (^{3}H)-Gamma-Aminobutyric Acid ([^{3}H]-GABA) Binding in the Human Brain : Alterations in Parkinson's Disease. Brain Res., 127, 269-278.

Lloyd, K. G., Dreksler, S. and Bird, E. D. (1977b). Alterations in ^{3}H-GABA Binding in Huntington's Chorea. Life Sci., 21, 747-754.

Lloyd, K. G., Worms, P., Scatton, B., Zivkovic, B. and Bartholini, G. (1979a). The influence of GABA on Dopamine Neuron Activity. In Presynaptic Receptors (eds S. Z. Langer, K. Starke and M. L. Dubocovich) Pergamon Press, New York.

Lloyd, K. G., Worms, P., Depoortere, H. and Bartholini, G. (1979b). Pharmacological Profile of SL 76 002, a New GABA-Mimetic Drug. In GABA-Neurotransmitters (eds P. Krogsgaard-Larsen, J. Scheel-Kruger and H. Kofod), Munskgaard, Copenhagen.

Lloyd, K. G., Munari, C., Worms, P., Bossi, L., Bancaud, J., Talairach, J. and Morselli, P. L. (1980a). The Role of GABA-Mediated Neurotransmission in Convulsive States. In GABA and Benzodiazepines Receptors (eds E. Costa, G. DiChiara and G. L. Gessa), Raven Press, New York.

Lloyd, K. G., Beaumont, K. and Ziegler, M. (1980b). Assessment of GABA Receptors in Different CNS Diseases by Means of Radioligand Assays. In Psychopharmacology and Biochemistry of Neurotransmitter Receptors (eds H. I. Yamamura, R. W. Olsen and E. Usdin), Elsevier, New York.

Mandel, P. and De Feudis, F. V. (1979). GABA-Biochemistry and CNS Functions. Plenum Press, New York.

Mandel, P. and De Feudis, F. V. (1980). Amino Acid Neuro-transmitters, Raven Press, New York.

McGeer, P. L. and McGeer, E. G. (1976). The GABA System and Function of the Basal Ganglia : Huntington's Disease. In GABA in Nervous System Function (eds E. Roberts, T. N. Chase and D. B. Tower) Raven Press, New York.

McGeer, P. L., McGeer, E. G. and Wada, J. A. (1971). Glutamic Acid Decarboxylase in Parkinson's Disease and Epilepsy. Neurology, 21, 1000-1007.

Morselli, P. L., Bossi, L., Henry, J. F., Zarifian, E. and Bartholini, G. (1980). In GABA and Other Inhibitory Neurotransmitters (eds S. Fielding and H. Lal), ANKHO, Fayetteville.

Olsen, R. W., Van Ness, P., Napias, C., Bergman, M. and Tourtellotte, R. W. (1980). GABA Receptor Binding and Endogenous Inhibitors in Normal Human Brain and Huntington's Disease. In Neurotransmitters and Peptide Hormones (eds M. J. Kuhar, S. J. Enna and G. Pepeu), Raven Press, New York.

Oppenheimer, D. R. (1976). Diseases of the Basal Ganglia, Cerebellum and Motor Neurons. In Greenfields Neuropathology, (eds W. A. Blackwood and J. A. N. Corsellis). Arnold, London.

Perry, T. L., Hansen, S., Berry, K., Mok, C. and Lesk, D. (1971). Free Amino Acids and Related Compounds in Biopsies of Human Brain. J. Neurochem., 18, 521-528.

Perry, T. L., Hansen, S. and Kloster, M. (1973). Huntington's Chorea : Deficiency of γ-Aminobutyric Acid in Brain. New Engl. J. Med., 288, 337-342.

Poirier, L. J., Sourkes, T. L. and Bédard, P. J. (1979). The Extrapyramidal System and Its Disorders, Raven Press, New York.

Reisine, T. D., Yamamura, H. T., Bird, E. D., Spokes, E. and Enna, S. J. (1978). Pre- and Postsynaptic Neurochemical Alterations in Alzheimer's Disease. Brain Res., 159, 477-481.

Reisine, T. D., Beaumont, K., Bird, E. D., Spokes, E. and
Yamamura, H. I. (1979). Huntington's Disease : Alterations in
Neurotransmitter Receptor Binding in the Human Brain. In
Huntington's Disease. Advances in Neurology, Vol. 23 (eds T. N.
Chase, N. S. Wexler and A. Barbeau), Raven Press, New York.

Ringel, S. P., Guthrie, M. and Klawans, H. L. (1973).
Current Treatment of Huntington's Chorea. In Huntington's Chorea:
1872-1972, Advances in Neurology, Vol. 1(eds A. Barbeau, T. N.
Chase and G. W. Paulson) Raven Press, New York.

Rinne, U. K., Sonninen, V., Riekkinen, P. and Laaksonen, H.
(1974) Dopaminergic Nervous Transmission in Parkinson's Disease.
Med. Biol., 52, 208-217.

Rinne, U. K., Koskinen, V., Laaksonen, H., Lönneberg, P. and
Sonninen, V. (1978). GABA Receptor Binding in the Parkinsonian
Brain. Life Sci., 22, 2225-2228.

Roberts, E. (1976). Disinhibition as an Organizing Principle
in the Nervous System - The Role of the GABA System. Application
to Neurologic and Psychiatric Disorders. In GABA in Nervous
System Function (eds E. Roberts, T. N. Chase and D. B. Tower),
Raven Press, New York.

Roberts, E., Chase. T. N. and Tower, D. B. (1976). GABA in
Nervous System Function, Raven Press, New York.

Scatton, B. and Bartholini, G. (1980). Modulation by GABA of
Cholinergic Transmission in Rat Brain. In GABA and Other Inhibi-
tory Neurotransmitters (eds S. Fielding and H. Lal), ANKHO,
Fayetteville.

Schwartz, W. J., Sharp, F. R., Gunn, R. H. and Evarts, E. V.
(1976). Lesions of Ascending Dopaminergic Pathways Decrease
Forebrain Glucose Uptake. Nature, 261, 155-157.

Scheel-Kruger, J. (1980). GABAergic Mechanisms in the Basal
Ganglia as Reduced from Local Cerebral Microinjections. In GABA
and Benzodiazepine Receptors (eds E. Costa, G. DiChiara and
G. L. Gessa), Raven Press, New York.

Shoulson, I., Chase, T. N., Roberts, E. and Van Balgooy,
J.N.A. (1975). Huntington's Disease : Treatment with Imidazole-
4-Acetic Acid. New Engl. J. Med., 293, 504-505.

Stahl, W.L. and Swanson, P.D. (1974). Biochemical abnormalities in Huntington's Disease. Neurology 24, 831-839.

Tower, D. B. (1976). GABA and Seizures : Clinical Correlates
in Man. In GABA in Nervous System Function (eds E. Roberts,
T.N. Chase and D. B. Tower), Raven Press, New York.

Worms, P., Depoortere, H. and Lloyd, K. G. (1979). Neurophar-
macological Spectrum of Muscimol. Life Sciences, 25, 607-614.

Wu, J. Y. (1976). Purification, Characterization and Kinetic
Studies of GAD and GABA-T from Mouse Brain. In GABA in Nervous
System Function (eds E. Roberts, T. N. Chase and D. B. Tower),
Raven Press, New York.

Yamamura, H. I., Olsen, R. W. and Usdin, E. (1980). Psychopharmacology and Biochemistry of Neurotransmitter Receptors,
Elsevier, New York.

Central Aminergic Function and its Disturbance by Hepatic Disease: The Current Status of L-valine Pharmacotherapy in Metabolic Coma

P. RIEDERER*, P. KRUZIK*, E. KIENZL*, G. KLEINBERGER†,
K. JELLINGER* and W. WESEMANN‡

* Ludwig Boltzmann Institute of Clinical Neurobiology,
  Lainz Hospital, A-1130 Vienna, Austria
† Dept. of Medicine I, University School of Medicine, A-1090
  Vienna, Austria
‡ Dept. Neurochemistry, Institute of Physiol. Chem. II,
  University Marburg/Lahn, West Germany

SUMMARY

Animal experiments and human post mortem brain analyses
led to the assumption that hepatic encephalopathy may
be more associated with disturbances of serotonin
(5-HT) metabolism than with disorders of catecholaminergic functions. Tryptophan (TRP), 5-HT, and 5-hydroxy
indole acetic acid (5-HIAA) have been found to be
substantially above normal values, especially in brain
stem areas. These disturbances in hepatic coma are
associated with high concentrations of ammonia and
glutamate. Both the high neuronal activity of 5-HT
and the high concentrations of the neurotoxic ammonia
may, therefore, contribute to the state of hepatic
coma. However, neuronal uptake of 5-HT is decreased
and 5-HT receptors appear to change their B_{max} and K_D
values with depth of coma. Reduced activity of
TRP-2,3-dioxygenase may be involved in stimulation
of 5-HT synthesis. Therefore, therapeutic strategies
which reduce both TRP and ammonia concentrations
should be employed. One possibility is the parenteral
administration of L-valine (VAL), a branched chain
amino acid. Its administration to patients with
liver cirrhosis reduced plasma ammonia which subsequently increased after VAL withdrawal. Patients
with higher baseline levels of ammonia responded
better than those with lower baseline ammonia. Such
treatment has been shown to decrease TRP, 5-HT, 5-HIAA
and ammonia levels in human brain tissue. Although

143

the mechanism by which VAL exerts its action is still
unknown, experimental evidence provides the following
possible explanations:
1. VAL competes with TRP at the blood brain-barrier;
2. VAL seems to generally decrease ammonia by inhibition of a) ammonia induced reduction of protein-
   synthesis or b) protein breakdown Utilization of
   VAL in the presence of high ammonia concentrations
   is, therefore, enhanced and induces lower plasma
   levels of branched chain amino acids but higher
   levels of aromatic amino acids.
3. Catabolism of VAL to products involved in energy
   metabolism.

INTRODUCTION

Experiments in animals have lead to the suggestion
that biogenic amines might be involved in metabolic
encephalopathies. Two theories have been established
in the last years.

First the hypothesis that disturbances in the
serotoninergic system contribute to hepatic encephalopathies is attractive for many reasons. In contrast
to tyrosine hydroxylase (TYR-OH),which under normal
conditions is saturated by its substrate tyrosine
(TYR), tryptophanhydroxylase (TRP-OH) is not fully
saturated by tryptophan (TRP). Increasing concentrations of TRP are found to enhance the rate of
production of cerebral serotonin (5-HT). Cerebral
TRP levels and, consequentially, neuronal uptake of
this precursor amino acid, are dependent upon the
concentration of the amino acid in blood and on the
exchange rate at the blood-brain barrier. Concentrations of aromatic amino acids including TRP are
increased in blood of patients with hepatic coma,
whereas branched chain amino acids are low (Record
et al. 1976, Cummings et al. 1976, Smith et al.1978,
Ferenci and Wewalka 1979). Factors which contribute
to an increase of TRP in the blood include disturbed
peripheral utilization of the amino acid due to
disturbed liver function and a drop in albumin
synthesis. The latter fact leads to increased
concentration of free TRP in plasma as under normal
circumstances about 9o% of the circulating TRP
is bound to this protein (Mc Menamy and Oncley,1958).
Therefore, in hepatic encephalopathies an increase

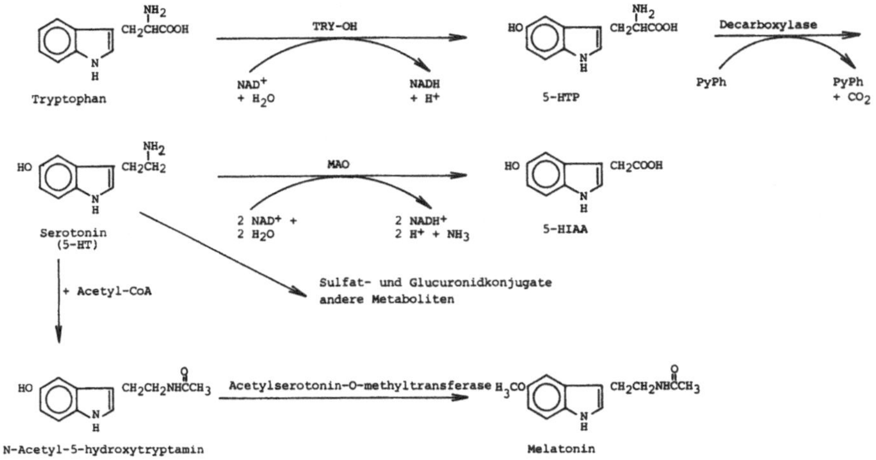

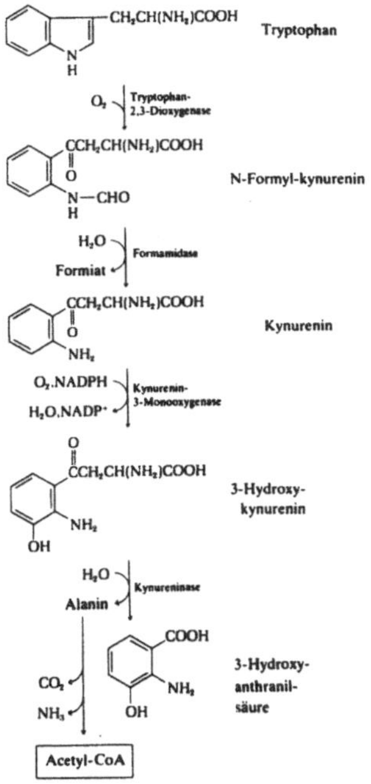

Figure 1. Important products in the metabolism of tryptophan

of peripheral TRP concentrations might enhance the cerebral and perhaps neuronal uptake of the amino acid thus leading to an increase in 5-HT synthesis and turnover (Fernstrom and Wurtman, 1972)(fig.1). Others have pointed out that increased formation of tryptamine as a result of increased TRP levels might play a role in the development of hepatic coma (Young and Lal, 1980). Also the role of other indolic TRP metabolites have been discussed in this respect (Sourkes, 1978).

The other theory is based on data showing that trace amines such as octopamine are substantially increased in the coma stage of animals with porto-caval shunts. Octopamine synthesized in the periphery can cross the blood-brain barrier and might therefore lead to a disturbance in the neuronal function. In fact, evidence from animal experiments assumes that octopamine replaces transmitterslike noradrenaline (NA) and might act as "false neurotransmitter" in this or other neuronal systems. A drop in the activity of cerebral TYR-OH might also be responsible for an increased synthesis of octopamine (Fischer et al.,1975).

PERIPHERAL DISTURBANCES IN THE AMINO ACID
POOL IN PATIENTS WITH LIVER CIRRHOSIS

Although there is no clear cut evidence that there exists a correlation between serum amino acid levels and the development of hepatic encephalopathy in liver cirrhosis with or without coma, some of the peripheral and central symptoms may indeed derive from the disturbed peripheral amino acid metabolism. As already mentioned, TRP seems to be one of the most interesting amino acids in this respect.

More than 9o% of TRP, however, are converted via kynurenine (KYN) and other metabolic products to nicotinamide. KYN is the first readily detectable metabolite in this major degradative pathway for TRP. The enzyme TRP-2,3-dioxygenase is the first step in the KYN pathway of TRP catabolism. KYN determination in physiological fluids after TRP load has been used as an indicator of in vivo TRP-2,3-dioxygenase activity in animals (Gal et al, 1977; Gould, 1979; Joseph et al., 1979). These recent studies demonstrated that KYN is present in rat and human brain and in view of

146

the demonstration of the necessary enzyme in rat
brain, make it likely that KYN is indeed synthesized
within the brain (Gal et al, 1977).

Agreeing with earlier reports (McMenamy and Oncley,
1958) we find that about 90% of plasma TRP is bound
to albumin (Gould,1979; Joseph and Kadam,1979)
(table 1)

TABLE 1.   Free and Total Tryptophan and Kynurenine
           in Plasma of Normal Subjects

	free	total	free % of total
TRP (μg/ml)	$1,83 \pm 0,14$ (8)	$18,8 \pm 5,3$ (14)	9,7
KYN (ng/ml)	$107 \pm 15,9$ (8)	416 ± 114 (14)	25,7

14 healthy controls (8 male, 6 female; age $36,5 \pm 7,4$ years)
Plasma samples were taken at 8.00 a.m.;fasting levels.
TRP was analysed according to Denckla and Dewey (1967),
KYN was determined by a GC method (Joseph and Risby,
1975). Free TRP and KYN were isolated according to
Riederer et al. (1975).

Patients with hepatic coma due to liver cirrhosis
(exogenous coma) and hepatic coma due to fulminant
failure (endogenous coma) have significantly higher
values of total TRP than controls or in liver
cirrhosis without coma (table 2).

The KYN synthesizing enzyme   (TRP-2,3-dioxygenase)
shows highest activity in the liver, but is also
active in other organs including brain tissue. Therefore it is of interest that in different stages of
hepatic coma serum KYN is significantly elevated
(Table 2). Moreover, an induction of the enzyme by
the increased TRP levels seems to be likely since
the increase in KYN-concentration is higher than that
for TRP.   The TRP/KYN ratio shows a linear decrease
correlating to the severity of hepatic failure. These
results are in agreement with findings by Hirayama
(1971), but are in contrast with other reports
(Record et al.,1976; Young et al.,1975; Cangiano et
al.,1976). The discrepancies can be explained by the
fact that all our patients hospitalized in an inten-

147

sive care ward were in the final stages of their illness and died in 5o% to 1oo% of cases depending on the severity of hepatic failure (Kleinberger et al., 1976). In contrast, all other studies mentioned were performed with less severely ill patients who had been hospitalized but could be discharged after successful drug treatment.

The reason for the high TRP values might be a rapid protein metabolism due to severe liver damage and/or diminished utilization of the amino acid. This assumption is supported by the findings that most other amino acids are also significantly higher compared to controls or liver cirrhosis without coma (Kleinberger,in preparation).

TABLE 2. Serum Tryptophan and Kynurenine Levels in Hepatic Coma

	total TRP μM	total KYN μM	$\frac{TRP}{KYN}$
1. Controls	$92^{\pm}25,9$ (14)	$2,0^{\pm}0,55$ (14)	46,0
2. Liver cirrhosis without coma	$135^{\pm}24,5$ (12)	$7,1^{\pm}1,1$ (12)[1]	21,8
3. Hepatic coma (liver cirrhosis)	$187^{\pm}15,7$ (14)[1]	$7,7^{\pm}1,8$ (14)[1]	17,8
4. Hepatic coma due to fulminant hep.failure	$237^{\pm}46,6$ (7)[1,2]	$17,4^{\pm}2,5$ (6)[1,2]	13,6

1) $p < 0,01$ compared to controls
2) $p < 0,01$ compared to hepatic cirrhosis without and with coma respectively
1.) 14 healthy controls (8 male,6 female; age:$36,5^{\pm}7,4$ years). These results were compared with a series of patients suffering from liver cirrhosis of different origin and stage.
2.) 12 patients suffering from liver cirrhosis without coma (9 male, 3 female; age: $52,4^{\pm}4,2$ years).7 pat. (58%) died in hospital. 3.) 14 pat. with hepatic coma due to liver cirrhosis (6 male,8 female; age: $49,2^{\pm}5,6$ years). 7 pat. (50%) died in hospital. 4.) 7 patients with hepatic coma due to fulminant hepatic failure (5 male, 2 female; age: $25,8^{\pm}4,5$ years). All of these patients died from acute virus hepatitis confirmed at autopsy.
TRP was determined according to Denckla and Dewey (1967), KYN according to Joseph et al. (1978).
TRP and KYN concentrations are fasting levels at 8:00 a.m.

The albumin binding of KYN is similar to that of TRP
(table 1). In hepatic encephalopathy an induction of
TRP-2,3-dioxygenase by the high TRP-plasma levels
could be determined indirectly by the estimation of
KYN (table 2). The pronounced increase of KYN in
comparison to the smaller increase of TRP could lead
to a displacement of TRP from albumin binding sites
thus   increasing the concentration of free TRP
(Young et al.; 1975). In fact, it has been suggested
(Tagliamonte et al., 1971) that the free form of TRP
is rather more important for brain uptake than the
amount which is bound. However, others point to the
view that total (free plus bound) TRP influences the
brain uptake of this amino acid (Fernstrom,1979;
Fernstrom and Wurtman, 1972). The increase in plasma TRP in
hepatic coma would not necessarily predict an increased
brain uptake of TRP as the competing amino acids,
phenylalanine and tyrosine, also increase several-fold
(Fernstrom and Wurtman,1972). However, fig.4,9 demonstrate that TRP is taken up to a large extent in
stages of severe hepatic coma. An explanation for
these data is given by experimental findings showing
that TRP is transported across the blood-brain
barrier (BBB) by two different uptake systems one of
which is a "low capacity uptake", which is saturable
and dependent on other aromatic and branched chain
amino acids, whereas the other is a "high capacity
uptake" only depending on blood TRP (Mans et al.,1979).
Shunt experiments show that during liver failure TRP
crosses the disturbed blood-brain barrier predominantly via the "high capacity" uptake (75%), independent
of competing amino acids (Mans et al.,1979). Whereas
this mechanism would explain the high TRP-values observed in brain of patients dying from such a disease,
the beneficial action of L-valine infusions in such
patients is perhaps better explained by a competing
action on the low capacity uptake of TRP. However, a
beneficial effect on the high capacity uptake cannot
be excluded, since we do not fully understand the
kinetics and influence of other amino acids on that
system. Therefore the determination of KYN, in the
cerebrospinal fluid and in brain tissue, might shed
light on the problem (see fig. 5 and table 4).

TABLE 3. Tryptophan Concentration of Various Areas of the Human Brain: Dependence on Post Mortem Time

Patients' diagnosis	age (years)	post mortem time (hrs.)	caudate N. µg/g	brain area raphé + R.F. µg/g	frontal cortex µg/g
7 MI, 8 BI, 1 MA 1 OC, 1 RC, 2 BP	$70,9\pm1,94$ (20)	$8,15\pm0,98$ (20)	$8,23\pm0,53$ (16)	$10,80\pm1,15$(17)	$8,35\pm0,64$(17)
n = 20					

$r = 0,47$ NS

$r = 0,90 \quad p<0,001$

$r = 0,44$ NS

Values are expressed as mean's $\pm$ s.e.m.; NS = not significant; r = linear regression-coefficient; MI = myocardial infarction; BI = brain infarction (analyses were performed on the contralateral site to the infarction); MA = morbus Alzheimer; OC = ovarial carcinoma; RC = rectal carcinoma; BP = broncho pneumonia.

Specimens of a variety of brain areas were obtained at autopsy. Post mortem time ranged from 3 to 20 hours. TRP was assayed according to Denckla and Dewey (1967). The design of such a procedure was chosen to differentiate between the metabolic situation in the post mortem brain and the time course of metabolic processes in excised brain samples (fig. 2)

THE CENTRAL AMINO ACID POOL IN
HEPATIC ENCEPHALOPATHY

A) The problem of amino acids analysis in autopsied
    human brain

The individual free amino acid levels in autopsy
human brain specimens vary over wide limits, but the
mean values are comparable within various laboratories
(Robinson and Williams, 1964; Perry et al., 1970;
Record et al., 1976; Weiser et al., 1978). Lajtha
and Toth (1974) investigated the post mortem changes
in amino acids occuring in brain autopsy specimens
of mice, rats and guinea pigs. Results were compared
between rapidly frozen brain, tissue kept at 20°C for
20 min. or 45 min. at 0°C and suggest that the post
mortem changes in the levels of most amino acids caused
by protein breakdown or by further metabolism are
rather small. Post mortem changes of amino acids in
human brain depend   on post mortem time, temperature
and oxygen. We have carefully studied the post mortem
changes of TRP and TYR. Both amino acids behave very
similar post mortem and so only the results with TRP
are given.

Figure 2 shows the increase in TRP content of human
brain tissue left at room temperature as well as at
2 - 4°C for varying time. In both cerebral cortex and
thalamus there is a substantial increase in TRP at
25°C which is significant from 5 to 22 hours post mortem.
At 2°C this increase is significantly less pronounced,
when compared to the results at 25°C. The increase
over time at 2 - 4°C seems to be more affected in the
thalamus than in the cerebral cortex. The reason for
this brain area specific difference seems to depend
on the varying enzyme activities due to the different
biochemical conditions. In this connection it is
noteworthy, that almost all patients who have to be
autopsied are kept at 2 - 4°C in cooling chambers at
the morgue. However, deep structures such as thalamus
may take longer time to reach 2 - 4°C than cerebral
cortex.

The post mortem increase at 25°C agrees with findings
by Young et al. (1977). However, when the brain
samples were flushed with N_2, TRP values were significantly lower in both brain areas, after 10 hours
storage, than those obtained with oxygen, demonstrating
the importance of an oxygen-free medium.

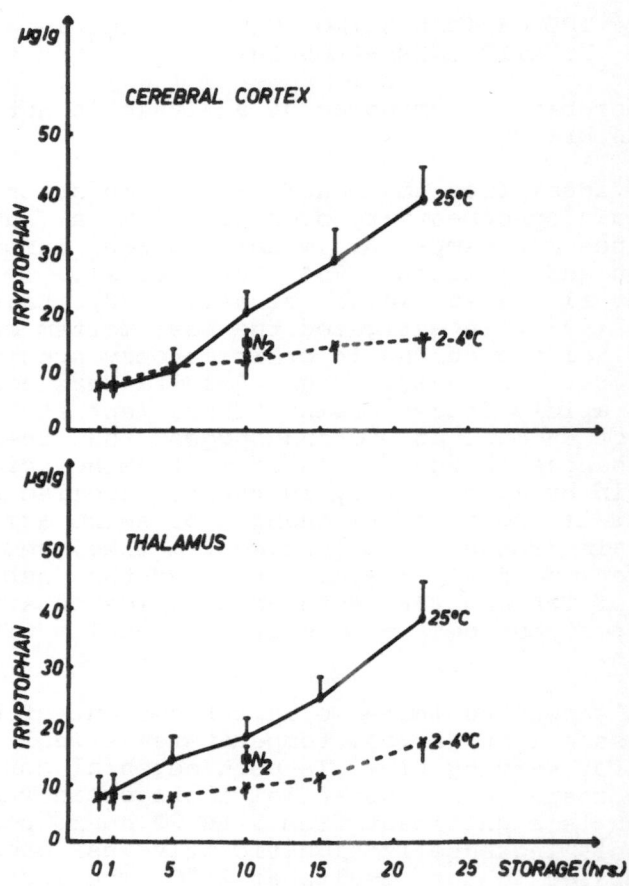

Figure 2.   Tryptophan in post mortem brains: dependence
on storage parameters

Specimens of cerebral cortex and thalamus were obtained from
4 patients (2 male, 2 female) who died of myocardial infarction
(age: 74 - 4 years). Autopsies were performed after 3 to
4 hours. After excision, tissue was divided into six pieces of
similar weight (150 mg) and left at room temperature (25°C)
for varying periods of time. The same procedure was carried
out at 2 - 4°C in a refrigerator. One experiment has been
performed by flushing the brain sample with dry nitrogen, in
order to examine the influence of oxygen. All samples were
kept in centrifuge tubes and these were wrapped with Parafilm.
Tissue was then frozen at -3o°C until analysis. TRP in this
study was measured by the method of Denckla et al. (1967). Values
(µg/g fresh weight) are means - s.d. from four independent
experiments.

Table 3 demonstrates that the metabolic processes in post mortem brains, which have not been excised for a varying length of time differ from these apparent in experiments shown in fig. 2.

In brain regions with a particularly high 5-HT metabolism, such as the raphe, we noted a highly significant increase of TRP with post mortem time. Cerebral cortex and caudate nucleus show a tendency, albeit not significant, for TRP to increase with time.

This finding underlines the importance of taking into account the post mortem time and condition of post mortem storage (temperature) when investigating amino acid precursors of neurotransmitters. In a recent study by Joseph et. al. (1979) this effect is clearly seen in the difference between cortex biopsy and post mortem samples (with a mean post mortem time of 51,4 hours), the latter showing an increase in TRP concentration greater than three fold (3,4 µg/g **vs** 12,1 µg/g respectively).

Thus, in such studies possible pathological differences may be masked by variation with post mortem time unless

a) shorter, more consistent post mortem times are taken, if at all possible, or
b) regression analysis of concentration vs. time is performed for both controls and pathological material.

The discrepancy between the first and the second experiment seems to depend on the content of oxygen (fig.2, table 3). Autolytic enzymes are included in those that have an oxygen-dependent activity (Lajtha, 1964). After death, the human brain is prevented from access to oxygen and therefore autolytic processes are slowed down. This may contribute to the relatively constant concentration of protein in post mortem brains (Rinne et al., 1973). In fact, we demonstrated that the content of protein in the cerebral cortex of the human brain post mortem is quite stable. As this is shown for an area in which TRY has also been measured, it is assumed that TRP, under optimal post mortem conditions, is not a major degradation product formed from brain protein. However, it is evident from these results that protein is partly degradated, possibly to amino acids, when temperature is higher (25°C) (Riederer et al., in preparation).

The results obtained show that these factors must be taken into consideration in studying tryptophan in human post mortem brain material. However, for post mortem times less than ten hours the slight increases are negligible.

B) Changes of amino acids in post mortem brains of patients with hepatic encephalopathy.

The amino acid data from human autopsy specimens of patients, comparable regarding patients' age and post mortem time, are shown in fig. 3. It is note-worthy that increased concentrations of glutamine, isoleucine and histidine were found in brains of patients with hepatic coma. Additionally, phosphoserine and methionine tended to higher values, while aspartate, phenylalanine, lysine, phosphoethanolamine and glycine tended to lower levels compared with control subjects. Furthermore, the content of ammonia was significantly elevated (data from Weiser et al.,1978). Valine treat-ment of patients with hepatic coma did not substantially change valine concentrations in the brain as might have been expected. This effect, together with the sig-nificant drop in glutamine and ammonia levels, suggest a connection between valine, ammonia detoxification and energy metabolism. In these treated patients aspartate, glycine, methionin, isoleucine, tyrosine, phenylalanine and histidinealso tended towards lower levels, whereas taurine, a presumably inhibitory amino acid neurotransmitter, was higher (fig.3). In control subjects and patients treated with L-valine, the brain amino acid differences between the various areas of brain examined were not significant. Only in patients with hepatic coma significant lower values were found for tyrosine and lysine in caudate nucleus compared with nucleus amygdalae (tyrosine: $0,16\pm0,02$ against $0,29\pm0,03$ μmol/g fresh weight; lysine: $0,17\pm0,03$ against $0,42\pm0,04$ μmol/g wet weight).

Kleinberger et al. (1977) reported that after complete parenteral nutrition and administration of 30-85 mg valine/kg/hour, the plasma valine level increased to 5990 μmol/l, but return to the initial level 24 hours after administration. Apparently, valine is partly removed from circulation by increased muscle uptake due to decreased hepatic insulin degradation. Smith et al. (1978) found that in dogs with an end-to-

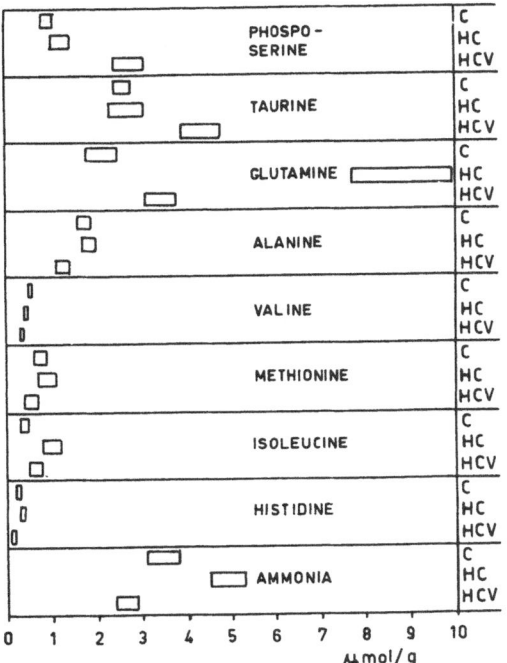

Figure 3   Alterations of brain amino acids in
            hepatic coma and hepatic coma after
            L-valine administration, respectively,
            compared with control subjects.

C control subjects, HC hepatic coma, HCV hepatic coma treated
with L-valine (5%; 25 ml/hour) and parenteral nutrition
(5o ml/hour). For further details see Weiser et al. (1978)·

side portocaval shunt, despite widely fluctuating but
decreased concentrations within the plasma, a somewhat
constant level of the branched-chain amino acids
within the cerebrospinal fluid is maintained. Inter-
estingly, decreased valine levels were detected in
the brain (table 6) though considerable amounts of this
amino acid were infused. Besides its competitive effects
on the aromatic amino acids and other neutral amino
acids and its increased muscle uptake (Fernstrom and
Wurtman, 1972), L-valine in brain could be metabolized
to α-oxo-glutarate via succinyl-SCoA and the tricarb-
oxylic acid cycle, thus supporting energy supplying
processes in brain and being available for detoxification of brain ammonia. Indeed we found significantly

155

lower brain ammonia levels after L-valine administration (table 6) and also the concentration of glutamine
was nearly normalized whereas that of glutamate was
decreased. To elucidate these results animal experiments are in preparation.

In control subjects and patients with hepatic coma
after L-valine administration we found no significant
differences in amino acids between the various regions
of brain. The reason for this may be the appreciable
variability in the content of the free amino acids
of each region. Thus it is not possible to distinguish
whether there are individual differences or differences between the various areas of the brain. However,
the significant differences in the content of tyrosine
and lysine in patients with hepatic coma between
caudate n. and n.amygdalae seem to demonstrate regional differences at least in hepatic coma.

C) Monoaminergic and related systems in metabolic
   encephalopathies

Recent studies suggest that disturbances in cerebral
neurotransmitter metabolism may be important in the
pathogenesis of hepatic encephalopathy and other
types of metabolic coma. In hepatic failure in both
animals and man a mild decrease of noradrenaline (NA)
and dopamine (DA) in the brain (Dodsworth et al.,1974;
Curzon et al., 1975), and increased brain and CSF
levels of tyrosine (TYR), tryptophan (TRP), serotonin
(5-HT), and 5-hydroxyindole acetic acid (5-HIAA)
have been found (Knell et al., 1974; Cummings et al.,
1976; Curzon et al., 1975; Young et al., 1975; Ono
et al.,1978; Bloxam and Curzon,1978; Jellinger et al.,
1978), suggesting an increase in 5-HT turnover in
hepatic coma. Similar changes of brain monoamines and
TRP have been recorded in both uremic and diabetic
coma in man (Jellinger and Riederer,1977,1978), and
in chronically uremic rats (Siassi et al.,1977), while
increased plasma and CSF TRP, and CSF 5-HIAA were reported in human subjects with uremic encephalopathy
(Sullivan et al., 1978). A characteristic amino acid
pattern in plasma, CSF and brain has been reported
in both experimental hepatic encephalopathies (Fischer
et al., 1976,1977; Smith et al.,1978) and human
patients with acute and chronic hepatic failure
(Record et al.,1976; Delafosse et al.,1977; Constantino

et al., 1978; Ono et al., 1978), with the greatest changes affecting the amino acids concerned with neurotransmitters in these conditions has been attributed, at least in part, to amino acid imbalance in plasma and brain (Fischer et al.,1976,1977). Experimental and clinical data indicate that the altered levels of aromatic and branched-chain amino acids in plasma and CSF correlate well with the development of hepatic encephalopathy (table 4) (Fischer et al.,1976,1977; Rosen et al., 1978; Smith et al., 1978), and that correction of the plasma amino acid abnormalities improves encephalopathy simultaneously with correction of neurotransmitter derangements in CSF (Smith et al.,1978) and brain (Jellinger et al., 1978), due to a shift in the competative brain uptake mechanisms of different amino acids.

TABLE 4. Tryptophan and kynurenine in human lumbar CSF

	controls	COMA hepatic	diabetic[2]
TRP µg/ml	1,5±0,3(15)	19,8±3,77 (14)[1,3]	2,7±0,53(11)
KYN ng/ml	37,0±8,0(5)	397,0±180 (14)[1,3]	54,0±23 (10)

$\bar{x}$ ± s.e.m.; number of patients in parentheses
1. Hepatic coma, stage III or IV according to Plum,1976 (patients were hospitalized in an acute ward and died as a consequence of liver cirrhosis within one week after lumbar CSF had been taken.
2. Insulin treated
3. p< 0,001 compared to controls

C)   The 5-HT system in metabolic encephalopathies

Studying the serotoninergic system, the following results with post mortem brain symples were obtained.

Tryptophan concentrations showed no definite changes
in non-comatose cases of liver cirrhosis, but were
significantly increased in hepatic coma, the highest
concentrations being found in the brainstem tegmentum,
lenticular nuclei, and some areas of the limbic system
(g.cinguli and amygdaloid nucleus). After L-valine
treatment of hepatic encephalopathy, TRP levels in all
brain regions approximated to control values (fig.4)

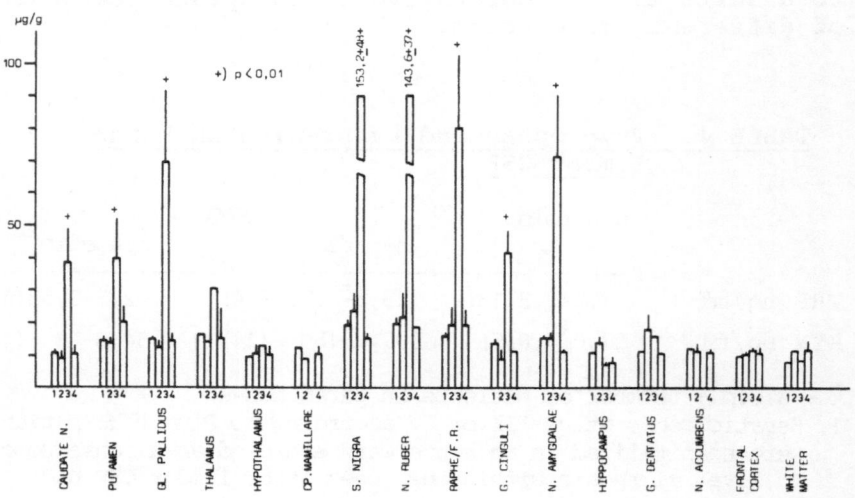

Figure 4   Post mortem human brain concentrations
             of tryptophan in metabolic encephalopathies

1:   controls
2:   hepatic failure without coma
3:   hepatic failure with coma
4:   hepatic coma treated with parenteral nutrition +L-valine

from Jellinger et al.,1978

158

Figure 5 demonstrates that KYN in controls is equally
distributed throughout the various brain areas. Its
distribution is similar to that of TRP (Birkmayer et
al.,1974) but differs from the well known brain area
dependent pattern of 5-HT (Birkmayer et al.,1974;
Bernheimer et al.,1961). Therefore, this finding
demonstrates a relationship of KYN with extraneuronal
tissue.

In hepatic coma the concentration of KYN is significantly increased in all brain areas. Its levels in
diabetic coma are between controls and hepatic coma
(Fig. 5). Hepatic coma treated with parenteral nutrition shows values which are in the control range
(fig. 5).

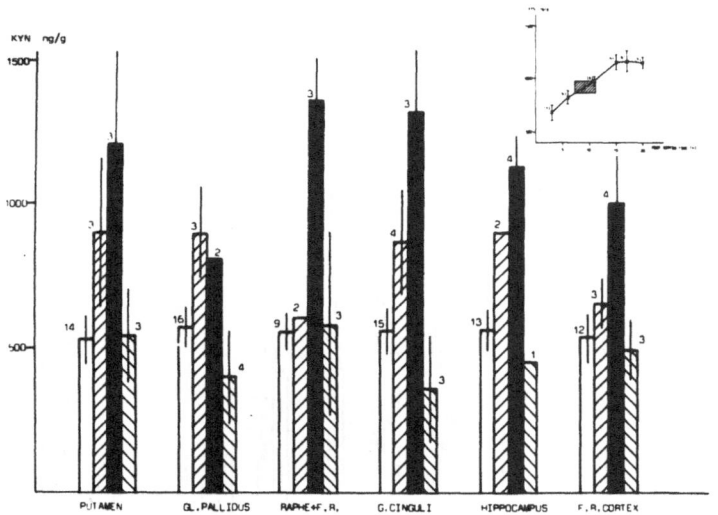

Figure 5   Post mortem human brain concentrations of
             kynurenine in metabolic encephalopathies

Number of patients in parenthesis;   mean's $\pm$ s.d. (ng/g fresh
weight).
x) insulin treated

☐ controls ▨ diabetic coma[x)] ■ hepatic coma (stage 3/4)
▨ hepatic coma - treated with L-valine

Insert to figure 5   Dependence of kynurenine concentration in
                      human post mortem brain tissue.
Values represent means's $\pm$ s.d. of 5 to 7 independent analysis of
brain tissue,composed of caudate n.,putamen,gl.pallidus,n.amygdala
and frontal cortex.Post mortem brain tissue of controls and pat.
                are indicated as shaded areas.

Serotonin was not significantly changed in liver cirrhosis without coma except for slight elevation in the
red nucleus, while a general increase was seen in
hepatic coma, the highest concentrations being found
in the brainstem tegmentum, lenticular nuclei and thalamus (fig.6). Although there were notable regional
differences in the individual brains, a similar increase
of 5-HT, particularly in the brainstem nuclei, was
obvious in both uremic and diabetic coma (Jellinger et
al.,1978). In hepatic encephalopathy treated with
L-valine, 5-HT levels in most brain areas were decreased below control values, the least severe reduction
being observed in the reticular formation (fig.6)-

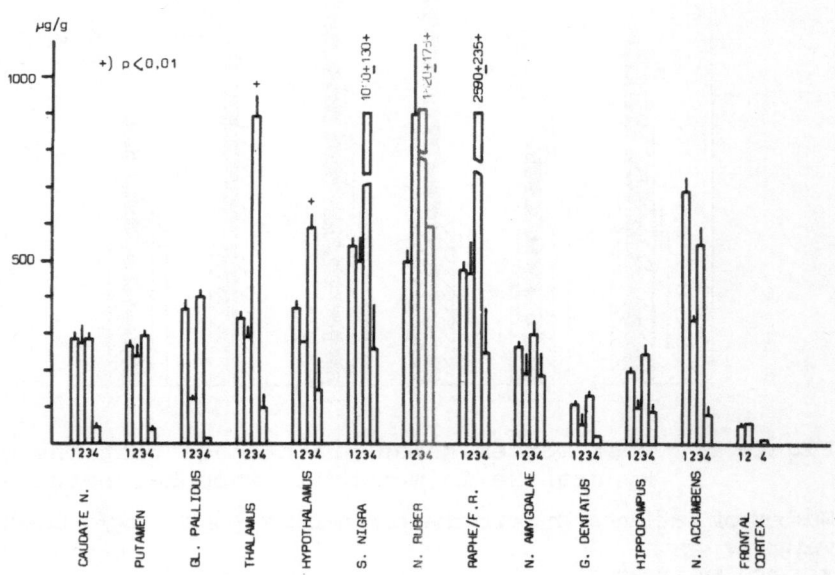

Figure 6   Post mortem human brain concentrations of
           serotonin in metabolic encephalopathies

1: controls
2: hepatic failure without coma
3: hepatic failure with coma
4: hepatic coma treated with parenteral nutrition + L-valine
from Jellinger et al.,1978

160

5-Hydroxyindole acetic acid showed no definite change
in liver cirrhosis without coma, but was significantly
increased in hepatic coma, the highest concentrations
being found in the lenticular nuclei, thalamus and
brainstem tegmentum (fig.7). A similar pattern was
observed in both diabetic and uremic coma, where the
highest increase in 5-HIAA was present in the brainstem tegmentum (fig.8). By contrast, hepatic encephalopathy treated with L-valine revealed 5-HIAA levels at
or even below control values (fig.7).

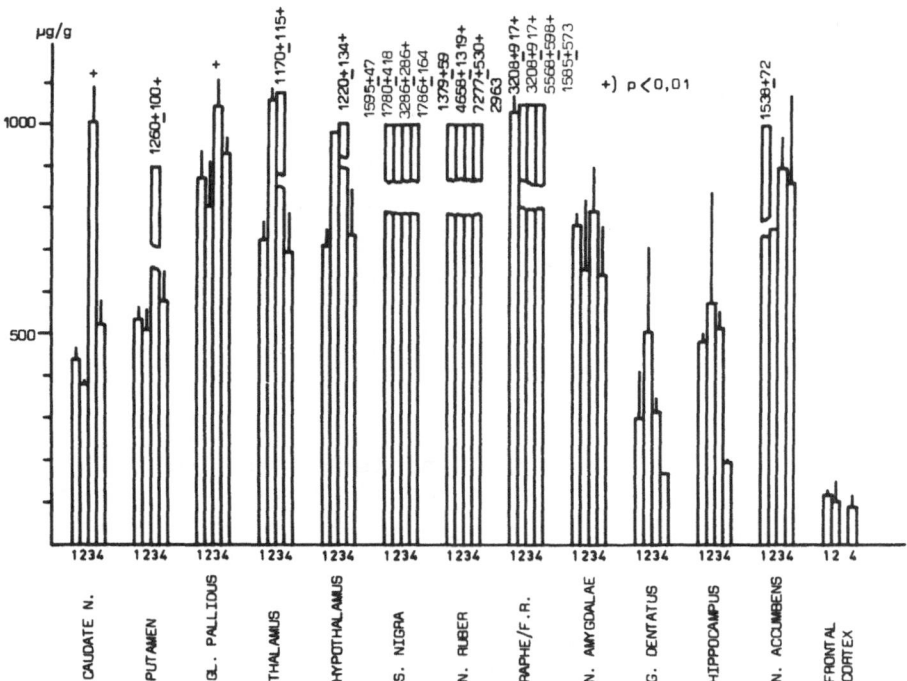

Figure 7   Post mortem human brain concentrations of
            5-hydroxyindole acetic acid in metabolic
            encephalopathies

1: controls
2: hepatic failure without coma
3: hepatic failure with coma
4: hepatic coma treated with parenteral nutrition + L-valine

from Jellinger et al.,1978

161

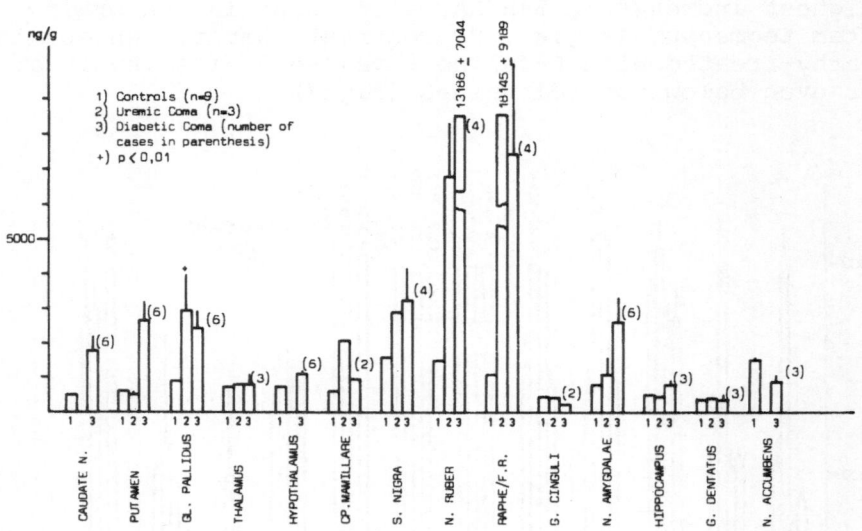

Figure 8   Brain 5-hydroxyindole acetic acid in
            controls and metabolic coma

As can be seen in fig.9 there is a slight but significant increase in brain TRP with increasing plasma
ratio when controls are plotted against liver cirrhosis without coma probably showing a rather intact
blood-brain barrier. These results are in close agreement with the experimental data (Fernstrom 1979). However, comatose patients show a sharp increase in brain
TRP when plasma ratio is not changed considerably in
comparison to cirrhosis without coma. This significant
increase of brain TRP might reflect disturbances in
the uptake of TRP across the BBB.

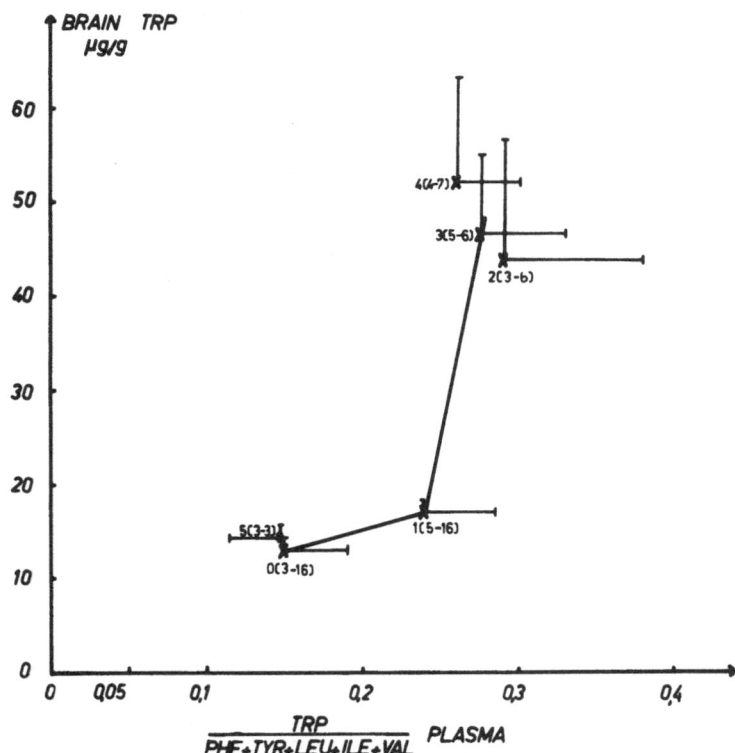

Figure 9   Dependence of human brain tryptophan
            concentrations on the ratio of serum
            tryptophan to competing amino acids
TRP = tryptophan   VAL = valine   LEU = leucine
ILEU = isoleucine   PHE = phenylalanine

0 = controls   1 = liver cirrhosis without coma
2 = exogenous hepatic coma   3 = endogenous hepatic coma
4 = fulminant hepatic failure (coma) due to viral infection
5 = hepatic coma treated with L-valine

First number in parenthesis gives number of brain-TRP values,
whereas second number gives the number of ratiodeterminations in
serum.   (Riederer et al.,1980)

As already mentioned (table 2) an increase of TRP in
the brain might be the result of an increased supply
deriving from the periphery due to disturbed liver
function. However, as more than 90 % of TRP is metabolized via the KYN-pathway, a reduced metabolism by
this metabolic rate could also contribute to an increased supply of TRP to serotoninergic neurons.

Brain KYN was significantly increased in hepatic coma
and less severely incrased in diabetic coma, while
after L-valine treatment of hepatic encephalopathy,
both TRP and KYN levels were about normal. These data
indicate that the cerebral KYN concentrations in metabolic encephalopathies parallels the changes in both
TRP and 5-HT in the brain. However, while in different
types of hepatic failure the TRP/KYN ratio in serum
shows a significant decrease, thus indicating an
induction of TRP-2,3-dioxygenase, in brain tissue
this ratio does not considerably change in hepatic
coma apart from in the raphe. These findings suggest
a fairly stable TRP turnover within the brain in
hepatic failure, and imply that in brain tissue only
a small metabolic induction by the 5-fold TRP is
possible. This finding is supported by recent experimental data indicating that KYN is synthetized from
TRP in the brain (Gal et al.,1977). Furthermore it has
been demonstrated that TRP, 5-HT and 5-HIAA are significantly elevated in human brain tissue of patients
with metabolic encephalopathies (Jellinger and Riederer,
1977; Jellinger et al., 1978).

A possible explanation for this effect is a decreased
metabolism of TRP through the KYN pathway. A drop in
the metabolism of TRP via KYN leads to an accumulation
of TRP which, after increased uptake in specific
serotoninergic neurons, might elevate 5-HT and its
turnover (Fernstrom and Wurtman, 1972.).

The rise of serum KYN in hepatic failure is similar
to the rise following a TRP load in normal subjects
(Joseph and Risby, 1975). While in normal brain tissue
the TRP/KYN ratio in controls is 20 - 30, in hepatic
and insulin-treated diabetic coma this ratio only
slightly differs, thereby demonstrating a different
behaviour of TRP and KYN in various types of metabolic
encephalopathies. It should be noted, however, that
part of the cerebral KYN may be derived from the
periphery, since KYN may penetrate into the brain via
the BBB (Gould, 1979; Joseph and Kadam,1979), thus
influencing brain TRP metabolism. However, no data
are available on the competing action of KYN on brain
uptake of TRP. Moreover, the uptake of KYN into brain
may be changed in patients with disturbed BBB due to
metabolic encephalopathies (Raskin and Fishman,1976;
Westergaard,1978).

Several studies in man as well as in animals have
shown that the amino acids, apart from TRP in both

plasma and CFS, are unchanged in cases with or without hepatic encephalopathy (Young et al.,1975; Record et al., 1976; Ono et al.,1978; Smith et al.,1978). These data and the present findings which show no significant changes in TRP, 5-HT and 5-HIAA values in human post - mortem brains of liver cirrhosis without coma suggest that in hepatic failure without clinical signs of encephalopathy there is no evidence for a severe derangement of brain indole metabolism.

By contrast, the results in human hepatic encephalopathy confirm previous findings in experimental hepatic coma, indicating, a significant elevation of brain TRP, and a regionally increased 5-HT level in the brainstem and striatum with uniform increase in 5-HT turnover as measured by the level of 5-HIAA (Curzon et al.,1975; Cummings et al.,1976). Although from the methodologi- cal point of view there may be some reservations concerning the comparability between experimental and human post-mortem studies in amino acids (Perry et al., 1970; Lajtha and Toth,1964), more recent results indicate that measurement of TRP and related compounds in post-mortem brain tissue may give satisfactory re- sults (Young et al.,1977).

Elevated brain levels of TRP in diabetic coma could be related to the insulin treatment of these patients, since high levels of circulating insulin are suggested to switch the balance between aromatic and branched - chain amino acids in favor of the latter ones (Soeters and Fischer,1976; Fernando et al.,1976; Kleinberger et al.,1977; Mackenzie and Trulson,1978). This could further explain the elevated brain levels of both 5-HT and 5-HIAA observed in diabetic coma. Some dis- orders in insulin and glucagon metabolism have further been suggested to occur in hepatic encephalopathies, and these may interfere with plasma amino acids uptake into the brain and thereby modify the synthesis and levels of brain monoamines (Munro et al.,1975; Soeters and Fischer,1976).

In both hepatic and renal insufficiency, there is an excessive accumulation in serum and CSF of free phenoles and phenolic acids (Record et al.,1976; Lesch,1973), and of free TRP (Gulyassi et al.,1970; Fischer et al., 1976; Curzon et al.,1975; Record et al.,1976; Sullivan et al.,1978; Bloxam and Curzon,1978) which may inter- fere with brain DA and 5-HT metabolism. As the changes of indoles in human brain do not considerably differ in hepatic, uremic, and diabetic coma, one may suggest some common derangement of cerebral monoamine neurotransmitter synthesis which, at least in hepatic failure, is attributed mainly to the imbalance of plasma amino acids competing for transport across the blood - brain barrier (Fischer et al.,1976,1977). In fact, it can be shown that the uptake of TRP from blood to brain changes depending on the severity of hepatic failure (fig.9). As a possible explanation, severe disturbances in BBB function (Raskin and Fishman, 1976, Westergaard,1978) with increased uptake of TRP may be responsible. Furthermore our data on brain TRP and KYN in such metabolic catastrophes give evidence for a disturbed metabolism of TRP-KYN pathway on the basis of reduced TRP-2,3-dioxygenase activity or stimulation. Although the plasma concentration of amino acids is suggested to be the main factor controlling their movement into the brain (Fernstrom and Wurtman,1972; Record et al.,1976; Young et al., 1975). the relationship between plasma and brain TRP, 5-HT and 5-HIAA appear to be very complex, and factors other than competitive uptake of amino acids may control brain DA and 5-HT synthesis (Heller, 1972; Hamon et al.,1974; Siassi et al.,1977; Bloxam and Curzon,1978).

Both hepatic and uremic encephalopathies are characterized morphologically by watery swelling of the astroglia and increased permeability of the BBB (Raskin and Fishman,1976; Laursen and Westergaard,1977) which have been related to increased brain 5-HT content, the pathophysiologic effects of this indole on the permeability of the BBB for proteins being well established (Westergaard, 1977). The fairly uniform accumulation of brain 5-HT in different types of metabolic coma suggests its basic importance for the development of brain edema, a common pathological feature in toxic and metabolic encephalopathies (Manz, 1976).

In both experimental hepatic encephalopathy and different types of human metabolic coma there is evidence for both general and regional increase of 5-HT levels and turnover, with preference for the brainstem tegmentum (raphe/reticular formation) and some parts of the limbic system, i.e. regions involved in the regulation of consciousness. Differences in the uptake mechanism of TRP between brain regions, notably in the raphe region, have been reported (Denizeau and Sourkes, 1977), and these may play a regulatory role in the

formation of 5-HT. There is considerable hevavioural
evidence that vigilance is regulated by the ascending
monoaminergic brainstem and mesolimbic systems, with
DA and NA having excitatory functions, while TRP and
5-HT tend to depress activity and to induce sleep
(Jouvet,1969; Morgane and Stern,1978). The marked
elevation in human metabolic encephalopathies of both
5-HT and 5-HIAA levels in the brainstem tegmentum and
in functionally connected parts of the hippocampus
suggest that increased 5-HT synthesis or turnover in
the ascending serotonergic brainstem-mesolimbic system
may represent an important biochemical basis of disorders of consciousness in metabolic catastrophes including hepatic coma (Jellinger and Riederer,1977,1978).

In fact, studies to further characterize the dynamic
changes of serotonergic neurons have shown that 5-HT
receptors change their kinetic behaviour. Table 5 gives
preliminary data regarding (high affinity) receptor
binding of the natural ligand to its receptor in
hepatic coma. For better comparison with data in the
literature a modification of the assay conditions
given by Bennett and Snyder (1976) have been chosen.
However, by going into more detail of the assay procedure (see Wesemann et al., this volume) further
changes might be recommended.
A trend to reduced B_{max} values and a drop in K_D is notable with this limited material. While the number of
receptors decreases, the affinity of serotonin to the
receptor apparently increases. A explanation of these
data in the light of increased serotonin synthesis and
turnover might be a reduced release of the transmitter,
thus leading to a "supersensitive" receptor. This
presumed mechanism might be the consequence of a reduced uptake of serotonin into vesicles. This speculation
could be valid because preliminary data regarding 5-HT
uptake show a diminuation of this process (Kruzik et
al.,1980). Two possible mechanisms might interfere with
the serotoninergic system and therefore block uptake
or receptor sites. Firstly, as mentioned above, a
number of phenolic and indolic compounds might interfere with basic neuronal processes. Secondly, it could
be shown in animal experiments, that the energy supply
of brain tissue might be reduced, although this is a
matter of controversy.

TABLE 5.   Serotonin Receptor Binding in Human Frontal Cortex Characterized by Scatchard Analysis

	age years	sex	"HIGH AFFINITY"		"LOW AFFINITY"	
			K_D nM	B_{Max} fmol/mg protein	K_D nM	B_{Max} fmol/mg protein
controls						
TM	81	F	11,9	380		
BM	80	F	4,4	228		
BR	66	M	5,7	242		
RA	86	F	4,2	159		
WK	71	F	4,65	222	40,0	550
precoma						
SJ	48	M	7,1	162		
KR	56	M	8,8	176		
coma hepaticum						
WR	61	M	1,7	212		
WR	51	M	4,2	65		
M.Parkinson						
KU	79	F	1,77	121	22,8	400
SJ	69	M	0,90	67	19,1	288

post mortem time was $5,3 \pm 1,13$ hours (all cases) (for further clinical details see Kienzl et al,1980) receptor analysis according to Bennett and Snyder,1976).

D)   L-VALINE AS A PHARMACOTHERAPEUTIC AGENT

Administration of L-valine leads to the accumulation of this branched-chain amino acid in plasma, but not in brain tissue (Kleinberger and Ferenci,1978; Weiser et al.,1978), while there is a significant decrease of aromatic amino acids in brain tissue and in serum as well as a decrease in brain ammonia (Jellinger et al.,1978, Weiser et al.,1978). These changes correlate well with both normalization of brain indole levels and rapid awakening of the patient form coma (Kleinberger et al.,1977, Kleinberger and Ferenci,1978). Since withdrawal of L-valine results in an increase of serum ammonia and clinical deterioration, the effect of this branched-chain amino acid in hepatic encephalopathy appears to be related to its competition with other amino acids and to an apparent ammonia detoxification (table 6)

TABLE 6.   The Influence of Parenteral Nutrition plus L-Valine on the Human Blood and Brain Ammonia Concentrations

	SERUM		BRAIN[B]	
	valine[A] μmol/ml	ammonia μg/100ml	valine μmol/g	ammonia μmol/g
controls	0,21±0,021	20 - 65	0,42±0,04(8)	3,2-3,8(8)
hepatic coma without PEN	0,24±0,056	266±87(8)[2]	0,37±0,04(6)	4,5-5,3(6)
hepatic coma +PEN+L-valine	2,17±0,35 [1,2]	145±25(8)[1,2] (235±32(6)[1,2] after L-valine withdrawal)	0,25±0,02(3)	2,4-2,8(3)[1]

$\bar{x}$ ± s.e.m.
PEN = parenteral nutrition (50 ml/hour) + 5% L-valine
1) = p < 0,01 when compared to hepatic coma without PEN
2) = p < 0,01 when compared to controls
A) = values taken from Kleinberger et al. (1977)
B) = data from Weiser et al.,1978
number of patients in parenthesis

There is experimental evidence that chronic ammonia loading results in significant incrase of pyruvate and lactate in both the brain and CSF (see Holm,1975; Fischer and Baldessarini,1976). In addition, there is a reduction of the Krebs cycle intermediate α-oxoglutarate and/or its turnover (see Holm,1975). Parenteral administration of amino acids without L-valine has no beneficial clinical action in hepatic encephalopathy (Kleinberger,pers.comm.). To explain the detoxification of ammonia by L-valine the following hypothesis has been proposed by Riederer et al., (1978) : L-valine is transaminated to α-oxo-isovaleric acid, which is oxidatively decarboxylated to isobutyryl-SCoA. This "active fatty acid" then is converted to methylmalonyl-SCoA by ß-oxidation, followed by conversion to succinyl-SCoA, which is an intermediary product of the citric acid cycle (fig.10).However, a stimulation of the citric acid cycle, resulting in a better availability and utilization of α-oxoglutarate, thus binding more ammonia, would only be decisive if this cycle is disturbed in metabolic encephalopathies. Experiments to prove or disprove this hypothesis for the action of L-valine in metabolic encephalopathies are underway in this laboratory.

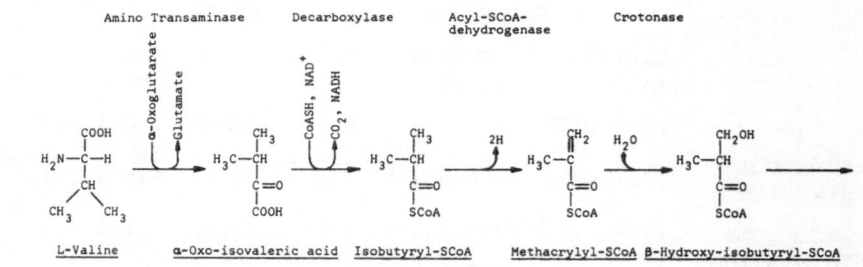

Figure 10   Interrelationship between L-valine and citric acid cycle

E)   THE CATECHOLAMINERGIC SYSTEM IN METABOLIC ENCEPHALOPATHY

Attention has been focused on catecholamines since they are involved in the "arousal" of animals and man. In hepatic encephalopathy catecholamines are reduced and consequently arousal is diminished. The loss of catecholaminergic activity has been taken into consideration to explain the state of coma by the "false neurotransmitter" theory (Fischer et al.,1975). How does this theory fit with human post mortem brain data?

1.   Tyrosine concentrations in the caudate nucleus showed a slight increase in both untreated and treated hepatic coma, with almost normal values in liver cirrhosis without encephalopathy. In contrast, a considerable increase was seen in diabetic coma (table 7).

2.   Tyrosine hydroxylase activity, examined in the caudate nucleus, revealed no significant changes in both hepatic and diabetic coma, but was considerably reduced in cases of liver cirrhosis and hepatic coma treated with L-valine who died from acute gastrointestinal hemorrhage (table 7)

TABLE 7.   Tyrosine and Tyrosine Hydroxylase Activity in Metabolic Comata

	Age (years)	Post mortem time (hours)	Tyrosine $\mu g/g$	Tyrosine hydroxylase activity nmoles/g.hours
controls (15)	$68,4 \pm 2,8$	$3,6 \pm 0,4$	$83,0 \pm 4,1$	$27,8 \pm 2,3$
coma hepaticum (4)	$52,0 \pm 4,8$	$15,8 \pm 7,4$	$109,3 \pm 9,2$	$25,2 \pm 9,1$
coma diabeticum (3)	$60,3 \pm 11,9$	$5,8 \pm 1,2$	$128,6 \pm 5,5$	$19,6 \pm 9,4$
liver cirrhosis without coma (3)	$74,3 \pm 7,0$	$4,5 \pm 0,3$	$93,2 \pm 2,0$	$10,0 \pm 0,7$
coma hepaticum +L-valine+P.N. (3)	$57 \pm 3,0$	$11,3 \pm 2,3$	$101,4 \pm 8,0$	$11,5 \pm 1,0$

number of patients in parenthesis
P.N. = parenteral nutrition
coma diabeticum treated with insulin

3. Dopamine was decreased in most brain regions of all types of metabolic coma, the average reduction being 20 to 30 percent of the controls. DA depletion was most pronounced in hepatic and uremic coma, with significant reduction in the corpus striatum, nucleus ruber, and nucl. accumbens (Jellinger et al.,1978) (fig.11).
In liver cirrhosis without encephalopathy adenylate cyclase in the caudate nucleus failed to be stimulated by 100 μM dopamine (table 8). This effect indicates a disturbance of the D_1-receptor function, which does not seem to be solely dependent on the reduced dopamine concentration. In vitro stimulation of the D_1-receptor unit should be possible under optimal assay condition unless failure in receptor coupling leads to a decrease of response. Thus a reduced activity of the D_1-receptor might contribute to reduced arousal reaction and coma. Whether endogenous toxins due to disturbed liver function contribute to inhibition of dopamine stimulation or reduced energy supply has to be clarified by further experiments.

TABLE 8   Stimulation of adenylate cyclase in caudate nucleus by dopamine

	cAMP (pmol/mg/min.)	
	Basal level	Dopamine stimulated (100 μM)
controls (6)	$54,2 \pm 7,3$	$90,3 \pm 16,5$
liver cirrhosis (5)	$65,1 \pm 13,8$	$45,2 \pm 8,9$

controls: 6 males; age: $70,3 \pm 5,42$ years; patients died by pulmonary emboly or heart failure; post mortem time $9,6 \pm 2,3$ hours

liver cirrhosis: 3 males, 2 females; age $69,5 \pm 5,2$ years; exogenous liver cirrhosis without encephalopathy; post mortem time $10,6 \pm 3,1$ hours

mean's $\pm$ s.e.m.; assay according to Kebabian et al. (1972) from: Riederer et al. (1978)

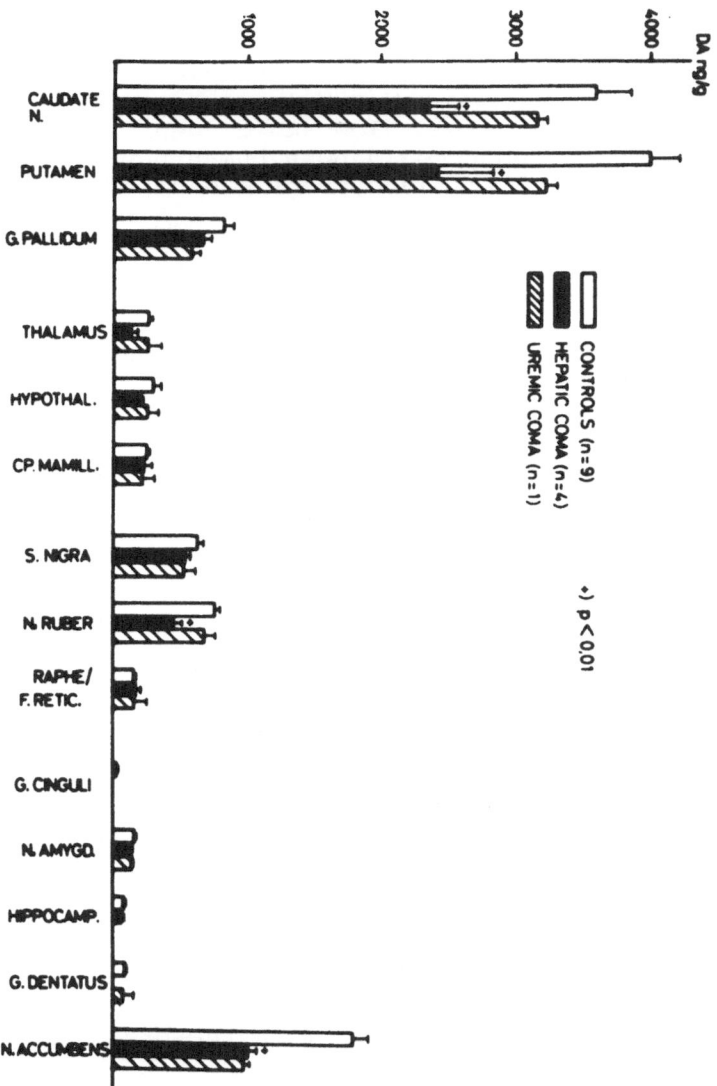

Figure 11   Brain dopamine in controls and metabolic
             coma

from: Riederer and Jellinger, 1979

However, our post mortem brain studies are not in line with the suggestion that "false neurotransmitters" like octopamine are involved in hepatic encephalopathies.

Decreased cerebral blood flow and oxygen uptake of the brain have been reported in both human hepatic coma (Fazekas et al.,1956; Zieve,1966) and experimental hepatic encephalopathy (Gjedde et al.,1978), and deficient energy supply of the brain has been interpreted as a major pathogenic factor of hepatic coma (Zieve,1966). Recent experimental studies, however, demonstrated that no shortage of cerebral energy supply exists during portocaval encephalopathy and hepatic coma (Fischer and Baldessarini,1976; Funovics and Fischer,1978). This assumption is confirmed by our preliminary findings in both hepatic and diabetic coma demonstrating no significant changes of TH activity in the caudate nucleus during these metabolic catastrophes (Jellinger et al.,1978).

The oxygen dependence of the TH (Fischer and Kaufman, 1972) suggests that brain oxygen tension should modify catecholamine synthesis in the brain, although decrease in TH synthesis caused by hypoxia can be prevented by pain or immobilization stress (Davis,1976). Simultaneous exposure to stress and hypoxia suggests that TH may undergo a conformational change in stressed animals resulting in increased affinity for oxygen (Carlsson, 1978). Whereas a decrease in rate of TH therefore does not necessarily accompany brain hypoxia, the activity of TH in ischemic gerbil hemispheres and striatum starts to decline significantly 12 hours after carotid ligation. This fall probably reflects a decrease in the quantity of enzyme protein secondary to impairment in its synthesis within perikarya (Moskowitz and Wurtman,1976). In essence, the direct inhibition of TH occurs already in moderate hypoxia and ischemia before any change in energy metabolism in the brain is detectable (Carlsson,1978).

The lack of decreased TH activity observed in human post mortem brain tissue from individuals with both hepatic and diabetic coma therefore is suggested to indicate a sufficient energy supply of the brain during such long-lasting metabolic catastrophes.

Evidence exists that neurons have a better affinity for oxygen during hypoxic stages. The result would be an enhanced catecholaminergic activity (Siesjö,1978). Therefore, long-term coma for some time does not alter the functional activity of O_2-dependent enzymes.

Clinical experiments support these findings. Long-term comatose stages are not necessarily accompanied or followed by neurological or psychiatric side effects.

From these and other experiments we suggest that an oxygen deficiency is primarily reflected by oxygen dependent enzymes of extraneuronal tissue, whereas the supply of oxygen, due to a higher affinity to neuronal systems is guaranteed over a relatively long period during metabolic catastrophes or hypoxia (Siesjö,1978, Davis and Carlsson,1972). Measurement of 5-HT and 5-HIAA again support the view that neuronal oxygen supply is guaranteed in metabolic encephalopathies (Jellinger et al.,1978; Joseph et al.,1978).

On the other hand, reduced TH activity in brain tissue from individuals with hepatic coma treated with L-valine who died from acute gastrointestinal hemorrhage is suggestive of severe ischemia of the brain due to this final bleeding.

As the oxygen-dependent TRP-2,3-dioxygenase presumably is located extraneuronally it is suggested that a deficiency of oxygen, which has been described in metabolic encephalopathies (Zieve,1966; Clavaria and Lascells,1976, Gjedde et al.,1978), could be reflected by a decrease in this enzyme activity. In fact, KYN is not substantially changed in brain tissue of patients with hepatic encephalopathies. Moreover, KYN concentrations led us to suggest that no induction of TRP-2,3-dioxygenase takes place in such metabolic disturbances. This could be the result of a lack in oxygen supply or inhibition by toxic products due to liver failure.

ACKNOWLEDGEMENT

This study was generously supported by a research grant from the Medical Scientific Fund of the Lord Mayor of Vienna and by Pfrimmer Pharmaceutics,Erlangen, GFR. The authors wish to express their gratitude to Mrs.A.Vigor for her excellent technical assistance and to Mrs.I.Riederer for secretarial work.

175

REFERENCES

Bennett, J.P., Snyder, S.H. (1975). Stereospecific binding of D-lysergic acid diethylamide (LSD) to brain membranes: Relationship to serotonin receptors. Brain Res. 94, 523-544.

Bernheimer, H., Birkmayer, W., Hornykiewicz, O. (1961). Verteilung des 5-Hydroxytryptamines (Serotonin) im Gehirn des Menschen und sein Verhalten bei Patienten mit Parkinson-Syndrom. Klin.Wschr. 39, 1056

Birkmayer, W., Danielczyk, W., Neumayer, E., Riederer P. (1974). Nucleus ruber and L-dopa psychosis: biochemical post mortem findings. J.Neural Transm. 35, 95-116

Bloxam, D.L., Curzon, G. (1978). A study of proposed determinants of brain tryptophan concentrations in rats after portocaval anastomosis and sham operation. J.Neurochem. 31, 1255-1263 .

Cangiano, C., Calcaterra, V., Cascino, A., Capocacci A. (1976). Bound and free tryptophan plasma levels in hepatic encephalopathy. Rendic.Gastroenterol. 8, 186-189.

Carlsson, A. (1978). Effects of low oxygen on brain monoamine metabolism. In Neurochemistry of Hypoxia, (ed.H.S.Bachelard), Proc.2nd Congr. Europ. Soc.for Neurochemistry, pp.266-270.

Clavaria, L.E., Lascells, R.L. (1976). In vitro polarographic studies in brain oxygen uptake in simulated uremic and hepatic coma. Acta neurol.scand. 54, 256-268.

Constantino, D., Borroni, M., Ambrosino, G., Martello, P., Rizzi,C., Olivotto, A.(1978). Variazioni degli aminoacidi plasmatici in alcumi casi di necrosie ed insufficienza epatica acuta nell' uomo. Min.Med. 69, 2355-2364.

Cummings, M.G., Soeters, P.B., James, J.H., Klane, J.M., Fischer, J.E. (1976). Regional brain indoleamine metabolism following chronic portocaval anastomosis in the rat. J.Neurochem. 27, 501-509.

Curzon, G., Kantamaneni, B.D., Fernando, J.C., Woods, M.S., Cavanagh, J.B. (1975). Effects of chronic portocaval anastomosis on brain tryptophan, tyrosine and 5-hydroxytryptamine. J.Neurochem. 24, 1065-1070.

Davis, J.N. (1976). Brain tyrosine hydroxylation: alteration of oxygen affinity in vivo by immobilization or electroshock in the rat. J.Neurochem. 27, 211-215 (1976).

Davis, J., Carlsson, A. (1972). The effect of hypoxia on mono-amine synthesis, levels and metabolism in rat brain. J.Neurochem. 19, 959-977

Delafosse, B., Bouletreau, P., Motin, J. (1977). Variation des acides aminés plasmatiques au cours de hepatites graves avec encéphalopathies. 10 observations. Nouv. Presse méd. 6, 1207-1211.

Denckla, W.D., Dewey, H.K. (1967). The determination of trypto-phan in plasma, liver and urine. J.Lab.Clin.Med. 69, 160-169.

Denizeau, F., Sourkes, T.L. (1977). Regional transport of tryp-tophan in rat brain. J.Neurochem. 28, 951-959.

Dodsworth, J.M., James, T.L., Cummings, M.C., Fischer, J.E.(1974). Depletion of brain norepinephrine in acute hepatic coma. Surgery 75, 811-820.

Fazekas, J.F., Howard, E.T., Ehrmanstraut, W.R., Alman, R.M.(1956). Cerebral metabolism in hepatic insufficiency. Amer.J.Med. 21, 843-849.

Ferenci, P., Wewalka, F. (1978). Plasma amino acids in hepatic encephalopathy. J.Neural Transm. Suppl.14, 87-94

Fernando, J.C.B., Knott, P.J., Curzon, G. (1976). The relevance of both plasma free tryptophan and insulin to rat brain trypto-phan concentration. J.Neurochem. 27, 343-345.

Fernstrom, J.D., Wurtman, R.F. (1972).Brain serotonin content: physiological regulation by plasma neutral amino acids. Science 178, 414-416.

Fernstrom J.D. (1979). Diet-induced changes in plasma amino acid pattern: Effects on the brain uptake of large neutral amino acids and on brain serotonin synthesis. J.Neural Transm., Suppl.15, 55-67.

Fischer, J.E., Baldessarini, R.J. (1976). Pathogenesis and therapy of hepatic coma. Prog. in Liver Dis. 5, 363-397.

Fischer, J.E., Funovics, J.M., Aguirre, A., James, J.H., Keane, J.M., Wesdorp R.I.C., Yoshimura, N., Westman, T. (1975). The role of plasma amino acids in hepatic encephalopathy. Surgery 78, 276-290.

Fischer, J.E., Funovics, J.M., Falcao, H.A., Wesdorp,R.I.C.(1976). L-Dopa in hepatic coma. Ann.Surg.183, 386-391.

Fischer J.E., Funovics, J.M., Aquirre, A., Wesdorp,R.I.C.,
James, J.H., Hodgman, J.K. (1977). Therapy of hepatic coma.
In Münchner Konferenz über neurologisch-psychiatrische Aspekte
des Komas, (eds. G. Dalle Ore, et al), Düsseldorf:Janssen,
pp. 223-239.

Fischer, D.B., Kaufmann, S. (1972). The inhibition of phenylalanine and tyrosine hydroxylase by high oxygen levels.
J.Neurochem.19, 1359-1365.

Funovics, J.M., Fischer, J.E. (1978). Brain energy metabolism
and alterations of transmitter profiles in acute hepatic coma.
J.Neural Transm., Suppl.14, 61-67.

Gál,E.M., Young,R.B., Sherman, A.D. (1977). Cerebral kynurenine
pathway: its significance to the metabolism of tryptophan.
6.Int.Meeting of the Intern.Soc.Neurochem.Copenhagen,Denmark
21.-26.8.1977   Abstract Nr. 521,p.565.

Gjedde, A., Lockwood, A., Duffy, T.E., Plum, F. (1978).
Cerebral blood flow and metabolism in chronically hyperammonemic
rats. Effect of acute ammonia challenge. Ann.Nerol.3, 325-330.

Gould S.E. (1979). The uptake of kynurenine, a tryptophan metabolite, into mouse brain. Meeting of the Brit.Pharmacol.Soc.
4.-6.4., Abstract C 74, pp 68-69.

Gulyassy, P.F., Aviram, A., Peters, J.H. (1970). Evaluation of
amino acid and protein requirements in chronic uremia.
Arch.Int.Med.126, 855-859.

Hamon, M., Bourgoin, S., Morot-Gaudry, Y., Hery, F., Glowinski,J.
(1974). Role of active transport of tryptophan in the control of
5-hydroxytryptamine synthesis. Adv.Biochem.Psychopharmac.11,153-162.

Heller, A. (1972). Neuronal control of brain serotonin. Fed.
Proc.31, 81-90.

Hirayama, C. (1971). Tryptophan metabolism in liver disease.
Clin.Chim.Acta 32, 191-197.

Holm, E. (1975). Ammoniak und hepatische Enzephalopathie.
    Biochemie, Elektrophysiologie, Toxikologie. Stuttgart,Fischer

Jellinger, K., Riederer, P. (1977). Brain monoamines in metabolic
(endotoxic) coma. J.Neural Transm. 41, 275-286.

Jellinger, K., Riederer, P. (1978). Brain monoamines in metabolic
coma and stroke. In Advances in Neurology, (eds. J.Cervos-Navarro
et al.), Vol.20, pp 535-546.

Jellinger, K., Riederer, P., Rausch,W.D., Kothbauer, P. (1978).
Brain monoamines in hepatic encephalopathy and other types of
metabolic coma. J.Neural Transm.Suppl. 14, 103-120.

Joseph, M.H., Baker, H.F., Lawson, A.M. (1978). Positive identification of kynurenine in rat and human brain. Biochem.Soc.
Trans. 6, 123-126.

Joseph, M.H., Risby, D. (1975). The determination of kynurenine
in plasma. Clin.Chim.Acta 63, 197-204.

Joseph, M.H., Kadam, B.V. (1979). Kynurenine: penetration to the
brain, effect on brain tryptophan and 5-hydroxytryptamine metabolism and binding to plasma albumin. Brit.J.Pharmacol.Meeting
of the   Brit.Pharmacol.Soc. 4.-6.4., Abstract C 73,pp 67-68.

Joseph, M.H., Baker, H.F., Crow, T.J., Riley, G.J., Risly, D.
(1979). Brain tryptophan metabolism in schizophrenia: a post
mortem study of metabolites on the serotonin and kynurenine
pathways in schizophrenic and control subjects. Psychopharmacol. 62, 279-285.

Jouvet, M. (1969). Biogenic amines and the states of sleep.
Science 163, 301-306.

Kebabian, J.W., Petzold, G.L., Greengard, P. (1972). Dopamine
sensitive adenylate cyclase on the caudate nucleus of rat brain
and its similarity to the "dopamine receptor". Proc.Nat.Acad .
Sci. (USA) 69, 2145-2149.

Kienzl,E.,Kruzik,P.,Riederer,P.,Arold,N.,Jellinger,K.,Wesemann,W.
(1980). Serotonin-receptor-binding und uptake-Studien bei hepa-
tischer Encephalopathie.Abstract:Tag.Ges.Biol.Chem,Innsbruck

Kleinberger, G., Ferenci, P. (1978). Die Beeinflussung der hepatischen Enzephalopathie durch L-Valin. In Ammoniak, Aminosäuren und
hepatische Enzephalopathie,(eds. F.Wewalka, B.Dragosics),
pp 207-214, G.Fischer, Stuttgart.

Kleinberger, G., Kotzaurek, R., Pall, H., Pichler, M., Szeless,
S. (1976). Parenterale Ernährung bei Coma hepaticum. Leber-
Magen-Darm 6, 340-346.

Kleinberger, G., Ferenci, P., Gassner, A., Lochs, H., Pall, H.,
Pichler, M. (1977). Behandlung des Coma hepaticum durch
vollständige parenterale Ernährung und L-Valin. Schweiz.med.
Wschr. 107, 1639.

Knell, A.J., Davidson, A.R., Kantamaneni, B.D.,Curzon, G.
(1974). Dopamine and serotonin metabolism in hepatic encephalopathy. Brit. Med. J. I, 549-551.

179

Kruzik P., Kienzl E., Riederer P. (1980). 5-Hydroxytryptamine uptake into human brain synaptosomes in controls and patients dying from hepatic encephalopathy. In preparation.

Lajtha, A. (1964). Protein metabolism of the nervous system. In Internat.Review of Neurobiology (eds. C.C.Pfeiffer, J.R. Smythies), Vol.6,pp 1-98, Academic Press, New York.

Lajtha, A., Toth, J. (1974). Post mortem changes in the cerebral free amino acid pool. Brain Res. 76, 546-551.

Laursen, H., Westergaard,E. (1977). The blood-brain barrier to horseradish peroxidase in rats with porto-systemic encephalopathy. Neuropath.appl.Neurobiol.3, 20-43.

Lesch,P. (1973). Toxische Metaboliten im Gehirn von Patienten mit Leberzirrhose und porto-cavalem Shunt. Dtsch.med.Wschr.98, 1929-1931.

Mackenzie, R.G., Trulson, M.E. (1978). Does insulin act directly on the brain to increase tryptophan levels? Neurochem.30,1206-1208.

Mans, A.M., Biebuyck, J.F., Saunders, S.J., Kirsch, R.E., Hawkins, R.A. (1979). Tryptophan transport across the blood-brain barrier during acute hepatic failure. J.Neurochem. 33,409-418.

Manz, H. (1976). The pathology of cerebral edema. Human Path.5, 291-313.

McMenamy, R.H., Oncley, J.T. (1958). The specific binding of 1-tryptophan to serum albumin. J.Biol.Chem.233, 1435-1447

Morgane, P.J., Stern, W.C. (1978). Serotonin in regulation of sleep. In Serotonin in Health and Disease, (ed.W.B.Hessman), Vol.2, pp 205-246. Spectrum Publ., New York.

Moskowitz, M.A., Wurtman, R.J. (1976). Acute stroke and brain monoamines. In Cerebrovascular Diseases, (ed. P.Scheinberg), pp 153-166, Raven Press, New York.

Munro, H.N., Fernstrom, J.D., Wurtman, R.J. (1975). Insulin, plasma amino acid imbalance, and hepatic coma. Lancet 1,722-724.

Ono, J., Mutson, D.G., Dombro, R.S., Levi, J.U., Livingstone,A., Zeppa, R. (1978). Tryptophan and hepatic coma. Gastroent.74, 196-200.

Perry, T.L., Hansen, S., Berry, K., Mok, C., Lesk, D. (1970). Free amino acids and related compounds in biopsies of human brain. J.Neurochem. 18, 521-528.

Plum, F., Hindfelt, B. (1976). The neurological complications of liver disease. In Handbook of Clinical Neurology, (eds. P.J. Vinken, G.W.Bruyn).,Vol.25,pp349-377,North Holland,Amsterdam-New-York.

Raskin, N.H., Fishman, R.A. (1976). Neurologic disorders in renal failure. New Engl.J.Med. 206, 204-209.

Record, C.O., Buxton, B., Chase, R.A., Curzon, G., Murray-Lyon, I.M., Williams, R. (1976). Plasma and brain amino acids in fulminant hepatic failure and their relationship to hepatic encephalopathy. Europ.J.Clin.Invest. 6, 387-394.

Riederer, P., Bamberg,E., Birkmayer, W. (1975). On the determination of free L-tryptophan. Acta vitamin.enzymol. 29, 21-24.

Riederer, P., Jellinger, K., Rausch, W.D., Kleinberger, G., Kothbauer, P. (1978). Zur Biochemie der hepatischen Enzephalopathien.Z.Gastroent. 16, 768-777.

Riederer, P., Jellinger, K. (1979). Brain monoamines in cerebral infarction and coma. In Pathophysiology of Cerebral Energy Metabolism, (eds.B.B.Mrsulja, Lj.M.Rakic, I.Klatzo,M.Spatz), pp 121-142, Plenum Publishing Corporation.

Riederer, P., Kleinberger, G., Jellinger, K. (1980). L-Valine: Mechanism of action on human brain in hepatic encephalopathy. 4.Int.Ammonia - Symp., Heidelberg,8.-11.5.

Rinne, U.K., Sonninen, V., Riekkinen, P., Laaksonen, H. (1973). Post mortem findings in Parkinsonian patients treated with L-DOPA: Biochemical considerations. In: Current Concepts in the Treatment of Parkinsonism, (ed.M.Yahr, Raven Press, New York), pp 211-233.

Robinson, N., Williams, C.B. (1964). Amino acids in human brain. Clin. Chim. Acta 12, 314-317.

Rosen, H.M., Yoshimura, N., Hodgman, J.M., Fischer, J.E. (1978). Plasma amino acid patterns in hepatic encephalopathy of differing etiology. Gastroenterology 72, 483-487.

Siassi, F., Wang, M., Kopple, J.D., Swendseid, M.E. (1977). Plasma tryptophan levels and brain serotonin metabolism in chronically uremic rats. J.Nutr.107, 840-845.

Siesjö, B.K. (1978). Brain energy metabolism and catecholaminergic activity in hypoxia, hypercapnia and ischemia. J.Neural Transm. Suppl. 14, 17-22.

Smith,A.R., Rossi-Fanell,F., Ziparo, V., James, J.H., Perelle, B.A., Fischer, J.E. (1978). Alterations in plasma and CSF amino acids, amines and metabolites in hepatic coma. Ann.Surg.187, 343-350.

Soeters, P.B., Fischer, J.E. (1976). Insulin, glucagon, amino acid imbalance and hepatic encephalopathy. Lancet 2,880-882.

Sourkes, T.L. (1978). Tryptophan in hepatic coma. J.Neural Transm. Suppl.14, 79-86.

Sullivan, P.A., Murnaghan, D., Callaghan, N., Kantamaneni,B.D., Curzon, G. (1978). Cerebral transmitter precursors and metabolites in advanced renal disease. J.Neurol.Neurosurg.Psychiat.41, 581-588.

Tagliamonte, A., Biggio, G., Gessa, G.L. (1971). Possible role of free plasma tryptophan in controlling brain tryptophan concentration. Rivista di Farmacologia e Terapia 2, 251-255.

Weiser, M., Riederer, P., Kleinberger, G. (1978). Human cerebral free amino acids in hepatic coma. J.Neural Transm. Suppl.14, 95-102.

Westergaard, E. (1977). The blood-brain barrier to horseradish peroxidase under normal and experimental conditions. Acta neuropath. (Berl.) 39, 181-186.

Westergaard, E. (1978). The effect of serotonin on the blood-brain barrier to protein. J.Neural Transm. Suppl. 14, 9-15.

Young, S.N., Lal, S., Sourkes, T.L., Feldmüller, F., Aronoff, A., Martin, J.B. (1975). Relationship between tryptophan in serum and CSF, and 5-hydroxyindoleacetic acid in CSF of man: effect of cirrhosis of liver and probenecid administration. J.Neurol. Neurosurg.Psychiat.38, 322-330.

Young, S.N., Tsang, D., Lal, S., Sourkes, T.L. (1977). Changes in the tryptophan content of excised human cerebral cortex. J.Neurochem. 28, 439-440.

Young, S.N., Lal, S. (1980). CNS Tryptamine metabolism in hepatic coma. J.Neural Transm. 47, 153-162.

Zieve, L. (1966). Pathogenesis of hepatic coma. Arch.int.Med.118, 211-233.

Antipsychotic Ligand Binding Sites in the Human Brain: Regional Selectivity Implicates Involvement of Multiple Neurotransmitters

ANNE C. ANDORN

Department of Psychiatry,
School of Medicine,
Case Western Reserve University,
Cleveland, Ohio, USA

Certain criteria should be satisfied to establish that a given drug exerts its effects by interactions with a specific neurotransmitter receptor (NTR). The first of these criteria is that the radiolabeled drug binding sites show selectivity similar to the known pharmacology of that NTR. Additional criteria include: 1) a correlation between ligand selectivity at drug binding sites and radiolabeled neurotransmitter sites, 2) evidence that the drug mimics or antagonizes an agonist mediated physiological event at that NTR, and 3) physicochemical identity of purified drug binding sites and the NTR. Based on these criteria, the investigation of the NTR involved in antipsychotic ligand action is in the preliminary stages.

Antipsychotic ligands have been shown to bind with high affinity at membranous sites that appear selective for dopaminergic, serotonergic and α-adrenergic agents in the central nervous system (CNS). It has therefore been speculated that the binding of antipsychotic ligands at apparent NTR correlates with the antipsychotic actions of these ligands or the pathophysiology of the psychoses. This correlation has appeared most direct for highly selective ligands or for diseases with well demarcated pathophysiology. In the case of less selective ligands or less well understood diseases, the correlation between events at the NTR and systemic events may be more remote.

Brief and even superficial review of the psychoses and antipsychotic ligand actions indicates that the disease may have muliple etiologies as well as a poorly understood pathophysiology, and that the ligands are not highly selective. The psychotic process itself is variable in expression. The varieties of psychotic illness are not easily demarcated, even by trained clinicians. Individual symptoms are difficult to objectively assess. Some symptoms, primarily those that can be seen in all psychoses (paranoid ideation, hallucinations) appear to respond to anti-

psychotic drugs. Other symptoms may persist even in the presence
of apparently adequate chemotherapy. There is, with present techniques, no microscopic or physiologic change that identifies psy-
chopathology in a given tissue. The antipsychotic ligands themselves exert multiple systemic actions (dopaminergic, α-adrenergic, histaminic and cholinergic antagonism) none of which is con-
clusively correlated with an antipsychotic effect. Antipsychotic
action is exerted only in the presence of symptoms of psychosis.
This action may even be reversed to a psychotic-like action in
non-psychotic individuals. In certain types of psychoses (deler-
ium/dementia) the psychotic symptoms appear to respond at lower
doses than do the same symptoms seen in the endogenous psychoses.
It therefore could be postulated that any observed interactions
between antipsychotic ligands and an NTR may only remotely correlate with either the systemic actions of the drugs or the pathophysiology of the diseases. However, a study of antipsychotic
ligand interaction with NTR may implicate sites of pathophysiology or modes of drug action previously unknown. Additionally,
if the CNS regions studied appear significant for psychotic
behavior in the human, then the correlation between events at the
NTR and systemic events might be more direct. Once putative NTR
are implicated in antipsychotic action an investigation of the
antipsychotic drug effects on agonist mediated events at these
NTR can then proceed.

METHODOLOGICAL COMPLICATIONS

Given the forgoing, it should not be unexpected that multiple high affinity binding sites for antipsychotic ligands
have been observed in the CNS (Table 1). Nor should it be unexpected that the selectivity of these binding sites should implicate the involvement of multiple NTR (Table 1). There is intra-
and inter-regional variation in the selectivity of sites observed. These binding sites have been demonstrated even in species
in which spontaneous endogenous psychoses have not been clearly
identified, as well as in normal human tissue.

Regional Variation in Selectivity
Previous studies of the caudate demonstrated that the selectivity of some antipsychotic binding sites appeared classically
dopaminergic. The selectivity also appeared to correlate with
the mg-potency ratio of antipsychotic drugs as recommended for
prescription in the human (Burt et al, 1976). However, no such
correlation for events known to be initiated by agonist binding
at dopamine receptors (eg. adenylate cyclase activation) could be
demonstrated (Iversen, 1975). α-Adrenergic selectivity has also
been observed at some antipsychotic ligand binding sites in rat
striatum (Andorn et al, in press). Antipsychotic binding
sites appear to exhibit serotonergic selectivity in frontal
cortex (Table 1). Antipsychotic drugs appear to compete

184

binding at dopaminergic, serotonergic and α-adrenergic NTR
when labelled with those respective ligands in CNS (Burt et
al, 1976; Peroutka, 1977; Peroutka, 1979). However, no
correlation between these events and the actions of the antipsychotic drugs is immediately evident. It appears therefore,
that the region studied selects for the NTR interaction
observed. The regional location of antipsychotic action is
unknown. Therefore, the NTR interaction observed in regions
studied to date may not be relevant for antipsychotic action.

TABLE 1. ^{3}H-Spiroperidol Binding in Brain

Species/Region	Number of Sites	K_D(nM) Range	B_{max} fmol/mg Range	NTR*	Ref.
Rat /Striat.	1or2	0.02–1.1.	100–500	DA,α–adr.	1–6
Calf /Caud.	1	0.30–0.36	227–363	DA	5,7
Human /Caud.	1	0.25–0.38	148–270	DA	1,5
Rat /F.C.	1or2	0.02–0.35	47–279	5HT	4,8,9
Calf /F.C.	NR	NR	NR	5HT	5
Human /F.C.	NR	NR	NR	5HT	10
Rat /Amygd.	NR	NR	NR	5HT	5
Calf /Amygd.	identified with ^{3}H–Haloperidol				11
Human /Amygd.	NR	NR	NR	5HT	5

The values recorded in this table represent the range
reported in the literature. Results reported below
are not included.Abbreviations used are: Striat.=stri-
atum; Caud.=caudate; F.C.=frontal cortex; amygd.=amyg-
dala; NR=not reported;DA=dopaminergic; α-adr.=α-adren-
ergic; 5HT=serotonergic; References are as follows:
1.Fields et al, 1977; 2.Howlett et al, 1978; 3.Leysen
et al, 1978; 4.Pedigo et al, 1978; 5.Creese et al,
1979; 6.Andorn et al, in press; 7.Creese et al, 1977;
8.Leysen et al, 1978; 9.Creese et al, 1978; 10.Mackay
et al, 1978; 11.Burt et al, 1976. Not included are
additional data demonstrating ^{3}H-spiroperidol binding
sites in many other brain regions, primarily the
limbic system (Howlett et al, 1979).

Species Specificity

High affinity binding sites for antipsychotic ligands have
also been demonstrated in similar regions of the human CNS,
suggesting that once regional binding sites are observed they
persist phylogenetically (Table 1). The preliminary studies in
these tissues indicate a similarity in the overall selectivity
of sites observed in the human and in the animal. However, the
demonstration of these binding sites in multiple species, as well
as in normal human tissue suggests that they may not be the sites
that are relevant for the uniquely human aspects of the psychoses.
It is possible that these sites may be dysfunctional in the psy-

185

chotic.patient, accounting for both the differential responses
to drugs and the disease.

Comparisons of antipsychotic binding site data in normal and
psychotic (usually schizophrenic) tissue is summarized in Table 2.

TABLE 2. Ligand Binding Studies in Psychotic CNS

Ligand	Region	Finding	Investigator
^{3}H-LSD	frontal cortex	$\uparrow B_{max}$	Bennet, 1979
^{3}H-5HT	frontal cortex	no change	Bennet, 1979
^{3}H-SPIRO	caudate	$\uparrow B_{max}$	Lee et al, 1978, 1980
^{3}H-SPIRO	caudate	" " ?K_D	Owen, 1979
^{3}H-HAL	caudate	$\uparrow B_{max}$	Lee et al, 1978, 1980
^{3}H-APO	caudate	no change	Lee et al, 1978
^{3}H-SPIRO	n. accumbens	no change	Mackay et al, 1978

Observations reported were made on human postmortem
tissue with variable postmortem intervals. In most
cases a single concentration of ligand was used to
compare total and specific binding in normal and
psychotic tissue. Psychotic tissue was identified
by chart diagnosis in all cases. Findings are the
only changes reported in reference to normal tissue.
Abbreviations:^{3}H-SPIRO=^{3}H-spiroperidol; ^{3}H-HAL=^{3}H-
haloperidol; ^{3}H-APO=^{3}H-apomorphine.

The most consistent observation has been an increase in the B_{max}
for ^{3}H-haloperidol or ^{3}H-spiroperidol specific binding , although
no such increase could be demonstrated for the dopamine agonist
apomorphine. At the ligand concentrations used in these studies
a small change in the B_{max} of very high affinity sites might be
obviated.

Binding Studies

It should be noted that in most instances an heterogeneity
of binding sites has been observed within a given region. Such
heterogeneity presents complications for the interpretation of
binding data, necessitating careful examination of each set of
sites. This is particularly relevant to the human studies in
which small tissue amounts necessitate examination at only one
ligand concentration. Further, competition studies need to be
performed under multiple conditions to selectively examine
different sets of sites.

Some investigators, as noted, observe an homogeneity of sites.
Methodological differences potentially account for these discrepancies. It appears that extensive hypotonic washing of the tis-

sue is requisite to demonstration of higher affinity sites(Andorn, in press). Because of these differences in tissue preparation and differences in binding assay procedure, comparison of studies done in different centers is limited. Comparison of studies using different retrieval systems for postmortem tissue is fraught with difficulty as noted elsewhere in this volume.

Although selectivity at antipsychotic ligand binding sites appears to be dopaminergic, serotonergic and α-adrenergic, no conclusive correlation can yet be drawn between these data and the interaction of antipsychotic ligands at NTR. Indications are that certain properties of some dopaminergic receptors in the caudate are altered in the psychoses.

FURTHER EVALUATION

As stated above, if specific NTR could be shown to consistently interact with antipsychotic ligands in CNS regions known to be associated with psychotic symptoms, then a correlation between the NTR and the disease or drug action might be more direct. The following studies were done to test the hypotheses that: 1) high affinity antipsychotic binding sites could be demonstrated in human CNS regions thought to be involved in the production of psychotic symptoms (based on previous lesioning studies), 2) these sites appear to have selectivity similar to pharmacologic NTR, 3) these sites appear similar in their selectivity to those identified in other regions of the human CNS: ie. they appear dopaminergic or serotonergic in pharmacologic selectivity.

In order to test these hypotheses, postmortem tissue was obtained at autopsy, prepared with extensive hypotonic washes, and studied with standard filtration binding assays employing ^{3}H-spiroperidol (^{3}HSP). Normal tissue was identified by the following criteria: 1) sudden death (accidental) without prolonged anoxia, 2) absence of known psychiatric or neurological disease, 3) absence of medications with psychotropic or sympathetic actions, 4) absence of substance abuse history (if questionable, drug screening was performed). In all cases, postmortem interval was less than 12 hours with an average of 6 hours. In all but one case the body was chilled immediately following death.

The areas chosen for study included: 1) prefrontal lobe, since the procedure of prefrontal lobotomy had met with some success in the alleviation of psychotic symptoms (Freeman, 1971), 2) the amygdala for similar results reported for bilateral amygdalotomies (Freeman et al, 1952), 3) the caudate as a reference tissue for comparison. Due to the amount of tissue available, the prefrontal area has been studied most thoroughly.

High affinity binding sites exhibiting saturability and specificity could be demonstrated in both prefrontal cortex and prefrontal white matter. In the prefrontal cortex, specific binding represented 75% of total binding, while in the prefrontal

187

white matter it represented less than 20% of total binding obviating further study. Saturation isotherms performed on prefrontal cortical gray homogenates were multiphasic and consistent with an heterogeneous population of sites (Figure 1). The approximate K_D

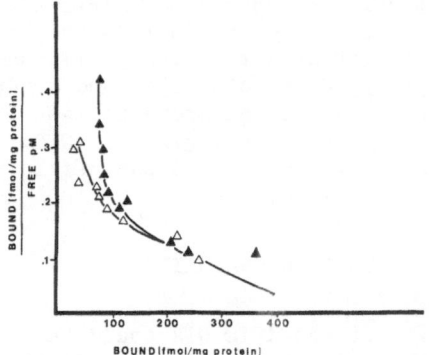

Figure 1. Saturation Isotherms of Prefrontal Cortex

Saturation isotherms were performed by placing increasing amounts of ^{3}HSP (0.01-4.0nM), membrane fragments(0.16mg/ml assay), and buffer (36 mM Hepes, pH 7.4 at 37°, 3.6 mM MgSO$_4$), and 0.1mM ascorbate in test tubes that were incubated for 30 min at 37° . The reaction was terminated by rapid filtration. Non-specific binding was defined as that in the presence of 100 μM (+)-butaclamol or haloperidol. These experiments were duplicated with each determination in triplicate on specimens of different postmortem intervals (▲ – 1 hr △ –12 hr) . Variation between duplicate experiments was less than 10%. If the Scatchard analysis is assumed biphasic only and the K_D are estimated from tangents drawn to the curve at intersection with the axes, then apparent K_D for the higher affinity site are 70 pM (▲) and 170 pM (△) and 1.5± 0.5 nM for the lower affinity site. B_{max} of the higher affinity site ranges from 80–120 fmol/mg protein and B_{max} of the lower affinity site is apparently 300± 20 fmol/mg protein.

ranged from 70±10 pM to 1.5 ± 0.5 nM . Replications of these studies on multiple samples indicated that the affinity of the highest affinity site was influenced by the postmortem interval (longer intervals, lower affinity) as shown in Figure 1.   The use of fresh membrane fragment preparations, those prepared from frozen homogenate or those that were frozen prior to use appears not to influence the saturation isotherms appreciably (data not

shown). Association and dissociation to prefrontal cortical
gray homogenates appears at least biphasic (Figure 2). The decay

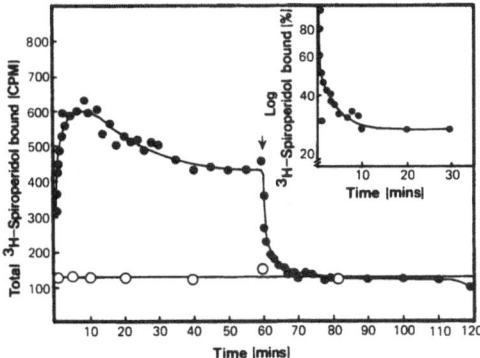

Figure 2. Association and dissociation to prefrontal cortex

  Association and dissociation were studied in conditions
  similar to Figure 1 with the exception that the incuba-
  ting mixture was placed in a flask and aliquots taken at
  the times indicated. Similar data was obtained in individual test tubes (not shown). This particular experiment (statistically indistinguishable from its duplicate) was performed at 1.0 nM [3]HSP on one hour post-
  mortem tissue. Similar findings have been observed at
  longer postmortem intervals. Non-specific binding
  (O) was defined as in Figure 1, and was linear over
  the time of observation. Dissociation was initiated
  by the addition of 100μM (+)-butaclamol.

noted in specific binding over time appears unrelated to protein
degradation, protein concentration, or ligand degradation (data
not shown). Studies to investigate ligand or receptor degradation during binding are current. It is possible to perform equi-
librium binding studies at 4 min or at 30+ min. Selectivity
studies performed at different ligand concentrations and incubation times in order to bias observations of different sets of
sites are summarized in Tables 3 and 4 . In general it appears
that antagonists are more potent competitors than agonists and
that α_1-adrenergic- and serotonergic-like selectivity are most
pronounced. Even though apomorphine is competitive at some sites
dopamine appears to be impotent at any [3]HSP concentration studied.
These data further indicate that there is a separation of α-adrenergic and serotonergic selectivity. Saturation isotherms performed with 100μM 5HT (Figure 3) or 10μM prazosin (not shown)
defining non-specific binding confirm this suspicion. 5HT
appears to compete only one set of sites labeled by [3]HSP in the
prefrontal cortex (Figure 3), while prazosin appears to compete

TABLE 3. Selectivity of Prefrontal Sites at 0.8nM ^{3}HSP

Drug	%Inhibition of Specific Binding	IC_{50} (nM)
Butyrophenones	100	1-40
Chlorpromazine	100	5
Cis-thiothixene	95	50
Trans-thiothixene	95	100
Molindone	80	1000
Prazosin	100	10
Phentolamine	100	500
Methysergide	70	100
Apomorphine	80	300
Dopamine	10	NA
Clonidine	80	$3 \cdot 10^4$
(-)-Norepinephrine	60	$3 \cdot 10^4$
5HT	60	1000
Tryptamine	60	10^4

Studies were done as in Figure 1 with the exception that
increasing concentrations of competitor were added to a
fixed concentration of ^{3}HSP (0.8nM). IC_{50} were determined
graphically as the concentration of competitor necessary
to inhibit 50% of specific ^{3}HSP binding as defined previously ., Experiments were usually performed in triplicate
and each result is the mean of three experiments on 1-3
hr. postmortem samples. At least six concentrations of
competitor were studied. NA means not applicable.

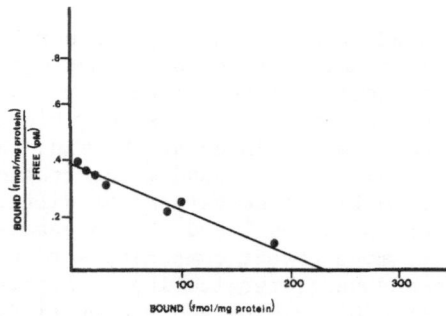

Figure 3. Saturation isotherm against 5HT

Saturation isotherms were performed as stated in
Figure 1 except that 100μM 5HT defined non-specific
binding and 4 min incubation times were used. Apparent K_D is 0.56 nM and B_{max} is 230 fmol/mg protein.
Replicates are statistically the same as this study.

190

TABLE 4. Selectivity in Prefrontal Cortex

Drug	0.08 nM ^{3}HSP IC_{50}(nM)	1.8nM ^{3}HSP IC_{50}(nM)
Butyrophenones	0.5–30	40–600
Chlorpromazine	3	100
Cis-thiothixene	30	10^4
Trans-thiothixene	0.25*	NA
Molindone	400	NA
Prazosin	3.2	200
Phentolamine	250	4000
Methysergide	2000	10^4
Apomorphine	300	NA
Clonidine	3000	NA
(–)-Norepinephrine	$3 \cdot 10^4$	10^6
5HT	$2 \cdot 10^4$	10^5
Tryptamine	10^5	NA
Dopamine	NA	NA
ACTH	NA	
Met-enkephalin	NA	
Substance P	NA	

Studies were performed as stated in Table 3 with the exception that an incubation time of 4 min was used for studies of 0.08 nM ^{3}HSP and 30 min was used for studies of 1.8 nM ^{3}HSP to bias for observation of very high and mid to low affinity sites respectively. All experiments were replicated but all points were done only in duplicate (* represents a single observation). Stereospecificity for (+)- and (–)-butaclamol is persistent in all conditions at greater than three orders of magnitude.

at least two sets of sites, whose affinity may or may not be the same as those competed by 5HT. The observation of an additional site with K_D 0.5 nM suggests that ^{3}HSP possibly binds at three sets of sites in human prefrontal cortex.

As shown in Figure 4, saturation isotherms of ^{3}HSP binding to human amygdala are similarly multiphasic, as are the association studies (Figure 5). The decrement in specific binding is again noted, although the time course at the same ligand concentration is slightly different (not shown). The evaluation of selectivity at these sites was begun under conditions to maximize observation of any mid and low affinity (K_D 0.4–1.0 nM) sites. In these conditions, the selectivity is very similar to that noted in the prefrontal cortex: α-adrenergic and serotonergic with absence of competition by dopamine. Dopamine

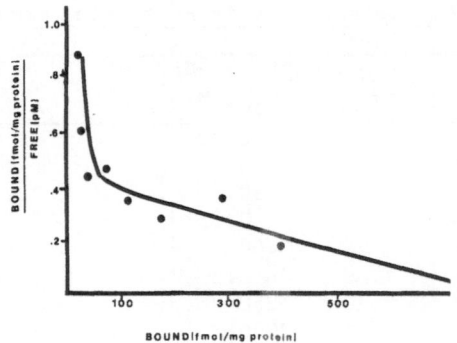

BOUND(fmol/mg protein)

Figure 4. Saturation Isotherms of Amygdala

Saturation analyses were performed as stated in Figure 1 with 4 min incubation on 1 hr postmortem tissue at a concentration of 0.085mg/ml assay. If the
Scatchard analysis is assumed biphasic, then the approximate K_D are 20pM and 1.5 nM with respective B_{max}
of 50 and 550 fmol/mg protein.

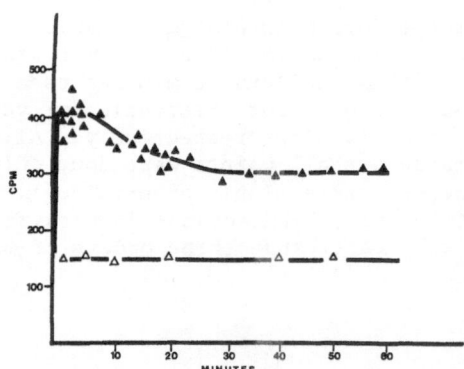

Figure 5. Association to Amygdala

These studies were performed as stated in Figure 2
at a final protein concentration of 0.085 mg/ml
assay. (▲) represents total binding and (△)
represents non-specific binding as defined in Figure 2.

competition has been assessed at several ligand concentrations
and incubation times.   The results indicate that dopamine does

192

not compete any ^{3}HSP binding present in human amygdala (not shown).

TABLE 5. Selectivity of Amygdalar Sites

Drug	% Inhibition of Specific Binding	IC_{50}(nM)
Butyrophenones	100	1-30
Fluphenazine	100	5
(+)-Butaclamol	100	30
(-)-Butaclamol	100	5000
Molindone	80	3000
WB4101	100	150
Phentolamine	95	500
Cinanserin	95	500
Methysergide	80	2000
Clonidine	90	$5 \cdot 10^4$
5HT	60	$3 \cdot 10^4$
Tryptamine	60	$3 \cdot 10^4$
(-)-Norepinephrine	50	10^4
(-)-Epinephrine	50	$3 \cdot 10^4$
Dopamine	20	NA

Competition studies were performed as previously stated with incubation time of 30 min at 25° and 0.5 nM ^{3}HSP. These conditions bias for observation of mid and low affinity sites (K_D 0.2-1.0nM) as shown by previous study (data not shown). A minimum of six concentrations of competitor were used for these determinations which were replicated in triplicate.

Observations at different incubation times and ^{3}HSP concentrations indicated a separation in selectivity for α-adrenergic and serotonergic agents. Saturation isotherms were performed on amygdalar membrane fragments using 100μM 5HT or 10μM WB4101 to define non-specific binding. As shown in Figure 6, and similar to the prefrontal cortex, Scatchard analyses of these data are biphasic for WB4101, but monophasic for 5HT. It appears from these data that WB4101 competes both high and low affinity sites while 5HT competes a mid affinity site (K_D 0.2 nM). However, this is speculative since exact determination of the K_D in multiphasic saturation isotherms is not possible.

Saturation isotherms of the caudate are again at least biphasic (Figure 7) with apparent K_D ranging from 20 pM to 1 nM in initial studies (three replications). Replicate studies of competition for ^{3}HSP binding in this region at two ligand concentrations (not shown) indicate that dopamine is a potent

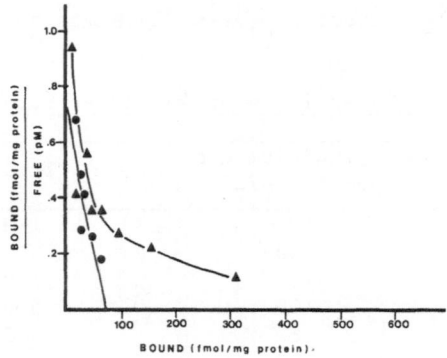

Figure 6. Saturation isotherms of amygdala:5HT and WB4101

Saturation isotherms performed using 100μM 5HT (●) or 10μM
WB4101 (▲) to define non-specific binding are shown. Conditions were the same as Figure 4. Approximate K_D for the
binding at sites competed by WB4101 are 20 pM and 1.0 nM
(with less than 10% variation between replicates). The
apparent K_D for sites competed by 5HT is 0.2 nM in
similar replicates.

competitor of ^{3}HSP binding (IC_{50} of 0.4-10 μM at various concentrations of ^{3}HSP). (-)-Norepinephrine may compete only a higher
affinity set of sites and serotonin is also a potent competitor
(IC_{50} of 30 μM) while methysergide competes with an IC_{50} of 1.0
μM at a ^{3}HSP concentration of 0.5 nM. These preliminary studies
do not indicate which site serotonergic agents compete. In general

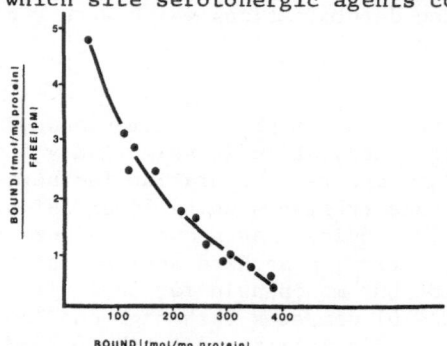

Figure 7. Saturation isotherms of caudate
Saturation studies were performed as stated (Figure 4)
on caudate homogenate. This experiment was replicated
with less than 20% variation. Approximate K_D are 20 pM
and 1.3 nM and respective B_{max} are 60 and 360 fmol/mg
protein for the two extremes.

overall selectivity appears similar to that observed previously in the rat (Andorn, in press).

Finally, limited tissue from a patient whose mental status exam by chart review met the research criteria for schizophrenia (Taylor et al, 1978) was obtained. This patient had not been known to be taking medication prior to his death. Toxicological screening at the time of autopsy was negative for antipsychotics, sedatives, sympathomimetics and drugs of abuse. The saturation isotherms of the prefrontal cortical homogenates of this patient were monophasic an apparent K_D of 0.78 nM (not shown). Preliminary selectivity studies showed similar changes in the affinity of competing ligands including antipsychotics and α-adrenergics. The saturation isotherms of the caudate were statistically indistinguishable from normal. The postmortem interval was felt to be slightly greater than 6 hours in this patient. The small amount of tissue has obviated sufficient study for conclusive interpretation.

DISCUSSION

^{3}H-Spiroperidol binding occurs with high affinity to an apparently heterogeneous population of sites (n=2 or 3) in human prefrontal cortex and amygdala. The sites appear to show selectivity for α-adrenergic and serotonergic agents but do not appear to demonstrate selectivity for dopamine. This selectivity appears to parallel the known pharmacological NTR if the contributing effects of multiple sites are considered in evaluation of the data. The competition by specific α-adrenergic agents for ^{3}HSP and ^{3}H-prazosin binding sites in prefrontal cortex appears correlative (Andorn, Mitrius, U'Prichard, preliminary observations) suggesting that spiroperidol binds at an α_1-adrenergic NTR in this region. As the prefrontal cortex and amygdala have been shown to have some significance for the production of psychotic symptoms, it is possible that dopaminergic NTR are not involved in the mediation of the psychoses. Observation that apomorphine is competitive for some sets of sites suggests that this agent is not a pure dopaminergic agonist but may have antagonist properties or act at another pharmacologic NTR.

The serotonergic selectivity demonstrated in these areas may not be representative of competition for serotonergic NTR. Although competition studies and saturation studies appear to indicate serotonergic interaction at a mid affinity site, it appears from preliminary studies (not shown) that serotonin has unusual actions on the dissociation of ^{3}HSP. Bennet (1978) observed that the B_{max} for ^{3}H-LSD sites (one of which is competed by spiroperidol) was decreased in schizophrenic frontal cortex, while that of 5HT sites was not. These observations taken together may suggest that 5HT is exerting remote(allosteric) effects on a common LSD-spiroperidol binding site that is not a serotonergic NTR.

The loss of specific binding over time as noted in association/dissociation studies of these regions could indicate that an endogenous ligand of nearly equal affinity to ^{3}HSP for the higher affinity site is still present, even in extensively washed preparations. Alternatively, these findings could indicate conformational changes or the presence of a high affinity site with rapid association and dissociation rates. It is also possible that a very high affinity site (K_D less than 10 pM) is unstable and is not being observed in saturation or competition studies. This is not unlikely given that these studies are performed at the limits of resolution of this system. Such an elusive site could also be dopaminergic.

It appears that ^{3}HSP labels an heterogeneous population of binding sites in human caudate. Further, it appears that while both sites exhibit dopaminergic selectivity, some sites also exhibit α-adrenergic selectivity. This is similar to previous observations in rat striatum (Andorn, in press). Taken together with previous animal studies, these findings suggest that binding sites demonstrated in the same region of several species may share a common overall selectivity, but may differ in other properties, as previously noted by Creese et al (1979).

Although the initial studies in schizophrenic tissue are hardly rigorous or complete, the data reported above indicate that functional differences in binding sites for ^{3}H-antipsychotic ligands can be observed between normal and psychotic individuals. These functional differences may account for differential responses to drug or for the disease.

The presence of binding sites with selectivity similar to pharmacologic NTR does not mean that antipsychotic action is mediated at these NTR. The changes observed in schizophrenic tissue do not mean that these NTR are dysfunctional in the psychoses. There is still no identified cellular or physiologic response modulated by these "antipsychotic receptors". Although speculation is tempting, extrapolation from these binding data to antipsychotic drug action or psychotic diseases is perhaps premature. It is apparent that although a final common pathway for antipsychotic drug action might be manifest in a single NTR (dopaminergic) the evidence does not favor an individual NTR as the single common denominator in antipsychotic drug action in health or disease.

ACKNOWLEDGEMENTS

The author is grateful for the advice and criticism of Dr. Michael E. Maguire, and the technical assistance of Linda E. Weber. Support for this work came from grants NIH RR 05410-18 and NINCDS - 1K07NS00468.

196

REFERENCES

Andorn, A.C., Maguire, M.E. ^{3}H -Spiroperidol Binding in Rat
  Striatum: Two High Affinity Sites of Differing Selecti-
  vities, J. Neurochem., in press.

Bennet, J.P., Enna, S.J., Bylund,D.B., Gillin,J.C., Wyatt, R.J.,
  Snyder, S.H. (1979). Neurotransmitter Receptors in Frontal
  Cortex of Schizophrenics. Arch. Gen. Psychiat., 36, 927-934.

Burt, D.R., Creese, I., Snyder, S.H. (1976). Properties of {^{3}H}-
  Haloperidol and {^{3}H}Dopamine Binding Associated with Dopamine Receptors in Calf Brain. Mol. Pharm., 12, 800-812.

Creese, I., Schneider, R., Snyder, S.H. (1977).^{3}H-Spiroperidol
  Labels Dopamine Receptors in Pituitary and Brain. Eur. J.
  Pharmacol., 46, 337-381.

Creese, I., Snyder, S.H. (1978). ^{3}H-Spiroperidol Labels Serotonin Receptors in Rat Cerebral Cortex and Hippocampus. Eur.
  J. Pharmacol., 49, 201-202.

Creese, I., Stewart, K., Snyder, S.H. (1979). Species Variations
  in Dopamine Receptors Binding. Eur. J. Pharmacol., 60,55-66.

Fields, J.Z., Reisine, T.D., Yamamura, H.I. (1977). Biochemical
  Demonstration of Dopaminergic Receptors in Rat and Human
  Brain Using {^{3}H}Spiroperidol. Brain Res., 136, 578-584.

Freeman, W., Williams, J.M. (1952). Human Sonar: The Amygdaloid
  Nucleus in Relation to Auditory Hallucinations. J. Nerv.
  Ment. Dis., 116, 456-462.

Freeman, W. (1971). Frontal Lobotomy in Early Schizophrenia,
  Long Followup in 415 Cases. Br. J. Psychiat.,119, 621-624.

Howlett, D.R., Nahorski, S.R. (1978). A Comparative Study of
  {^{3}H}Haloperidol and {^{3}H}Spiroperidol Binding to Receptors on
  Rat Cerebral Membranes. FEBS Letts., 87, 152-156.

Howlett, D.R., Morris, H., Nahorski, S.R. (1979). Anamolous Properties of {^{3}H} Spiperone Binding Sites in Various Areas of
  the Rat Limbic System. Mol. Pharmacol., 15, 506-514.

Iversen, L. (1975). Dopamine Receptors in Brain. Science, 188,
  1084-1089.

Lee,T., Seeman, P.(1978). Binding of ^{3}H-Neuroleptics and ^{3}H-Apo-
  morphine in Schizophrenic Brains. Nature, 274, 897-900.

Lee, T., Seeman, P. (1980). Elevation of Brain Neuroleptic/Dopa-
  mine Receptors in Schizophrenia. Am. J. Psychiat.,137,190-
  197.

Leysen, J.E., Gommeren, W., Laduron, P.M. (1978). Spiperone: a Ligand of Choice for Neuroleptic Receptors. Biochem. Pharmacol., $\underline{27}$, 307-316.

Leysen, J.E., Niemegeers, C.J.E., Tollenaere, J.P., Laduron,P.M. (1978). Serotonergic Component of Neuroleptic Receptors. Nature, $\underline{272}$, 168-171.

Mackay, A.V., Doble, A., Bird,E.,Spokes,E.G.,Quik,M., Iversen, L.L. (1978). ^{3}H-Spiperone Binding in Normal and Schizophrenic Post-Mortem Human Brain. Life Sci.,$\underline{23}$, 527-532.

Owen,F.,Crow,T.J.,Poulter,M.,Cross, A.J., Longden,A.,Riley,G.J. (1978). Increased Dopamine Receptor Sensitivity in Schizophrenia. Lancet,   224-226.

Pedigo,N.,Reisine, T.D., Fields, J.Z., Yamamura, H.I. (1978). ^{3}H-Spiroperidol Binding to Two Receptor Sites in Both the Corpus Striatum and Frontal Cortex of Rat Brain. Eur. J. Pharmacol., $\underline{50}$, 451-453.

Peroutka, S.J., U'Prichard, D.C., Greenberg, D.A., Snyder, S.H., (1977). Neuroleptic Drug Interactions with Norepinephrine Alpha Receptor Binding Sites in Rat Brain. Neuropharmacol.,$\underline{16}$, 549-556.

Peroutka, S.J. Snyder, S.H. (1979). Multiple Serotonin Receptors:Differential Binding of{^{3}H}5-Hydroxytryptamine,{^{3}H}-Lysergic Acid Diethylamide and{^{3}H}Spiroperidol. Mol. Pharmacol., $\underline{16}$, 687-699.

Taylor, M.A., Abrams, R. (1978). The Prevalence of Schizophrenia: A Reassessment Using Modern Diagnostic Driteria. Am. J. Psychiat., $\underline{135}$, 945-948.

Section V
Pathological States
(A) Huntington's Disease

Transmitter Biochemistry of Huntington's Disease

EDWARD D. BIRD and LINDA J. KRAUS

Ralph Lowell Laboratory,
McLean Hospital and
Massachusetts General Hospital,
Harvard Medical School,
Boston, Massachusetts, USA

A great deal of new data on the neurochemistry of human brain has resulted from the availability of post-mortem brain tissues from patients dying from Huntington's disease (H.D.).

This chapter will review some of this data, and discuss interpretations of the findings. A clinical description of this disorder will provide a basis for some of these interpretations.

Huntington's disease (H.D.) is an autosomal dominant inherited disorder of the nervous system. Each child of an H.D. affected parent has a 50% chance of inheriting the disorder (Oliver, 1970, Reed and Neel, 1959, Wallace, 1974, Wendt and Drohm, 1972). To date, no effective diagnostic test is available to determine which individuals among those genetically "at risk" for the disease do in fact carry the H.D. gene and will inherit the disorder (Manyam, Hare, Katz and Glaeser, 1978).

The disease, characterized by choreiform movements, will usually first appear as occasional involuntary movements of the extremities, with twitching of the face, progressing over a 15 to 20 year period to involve more muscle groups until the patient is unable to walk, stand, speak intelligibly, or even eat (Bruyn, 1968). The average age of onset is in the early forties, after the age of childbearing.

Progressive dementia is also a feature in most H.D. cases (Lieberman, Dziatolowski, Neophytides, Kupersmith, Aleksic, Serby, Viorein and Goldstein, 1979). Many patients will also manifest behavioral changes before the onset of movements (Folstein, Folstein and McHugh, 1979). As a result, these patients are often confined to a mental hospital with a diagnosis of schizophrenia, which is only corrected years later after the onset of movements. Though onset in the forties is most common, both a juvenile and a late onset form of the disease are also recognized. The juvenile form may include epilepsy, myoclonic or dystonic rather than choreiform movements, and often muscle rigidity is present. The course of the disease is more rapid (Westphal, 1905), whereas when the onset is in the fifth or sixth decade the progress of the disease is usually slower and there is very little sign of dementia. The progressive deterioration of H.D. may span 10 to 20 years and patients usually die of pneumonia or other complications of their debilitated state.

Huntington's disease is marked pathologically by the loss of neurons in the caudate nucleus and putamen (Alzheimer, 1911). While there is some atrophy of neurons throughout the brain, the greatest atrophy occurs in the basal ganglia. The whole brain weight is usually reduced to 20% of normal, but the weight of the basal ganglia is often less than 50% of normal. The striatal complex which contains the nucleus accumbens, a part of the limbic system, is less atrophic than the more posterior regions of the caudate and putamen (Bruyn, 1968). The brain stem is also atrophic, particularly in the region of the substantia nigra.

Biopsy of the frontal cortex reveals a proliferation and hypertrophy of astrocytes containing large amounts of lipofuscin and high acid phosphatase activity (Tellez-Nagel, Johnson and Terry, 1973). The lipofuscin granules though increased are not unusual in appearance. Increased amounts of smooth endoplasmic reticulum and vesicles associated with Golgi complexes are also observed (Tellez-Nagel et al, 1973).

On microscopic examination, tissue loss is most pronounced in the caudate, putamen, globus pallidus and substantia nigra. The concentration of glial cells in the basal ganglia is increased due to the loss of neuronal cells, though the total glial cell number has been shown to be the same in H.D. as normal brains (Lange, Thorner, Hopf and Schroder, 1976).

Given the hereditary aspect of H.D., it is of particular importance that cases diagnosed clinically as H.D. be confirmed histopathologically at autopsy. We have found that some 7% of cases clinically diagnosed as H.D. were in fact other neurologic conditions. Clearly, histopathologic examination of tissue to confirm the diagnosis of H.D. is also an essential pre-requisite to any research studies of these tissues.

The careful dissection of brain tissue obtained following prompt autopsy and preservation at -70°C is the cornerstone
of neurochemical studies on H.D. and other extrapyramidal disorders. Only through rigorous attention to the collection and
preservation of brain tissues from large numbers of patients
and standardization of dissection procedures to identify
smaller discrete areas will accurate neurochemical studies be
possible. In turn, these neurochemical studies may provide a
better understanding of the underlying neuropharmacologic basis
of disease and new hope in their management and treatment.
Although the specific nature of the gene defect responsible for
the premature death of specific nerve cells in Huntington's
disease is unknown, there is considerable evidence to suggest
that an imbalance of neurotransmitters in the brain, particularly γ-aminobutyric acid (GABA), dopamine, and acetylcholine,
may account in part for some clinical manifestations of the
disorder.

GABA, an important inhibitory neurotransmitter, is
widely distributed throughout the CNS and attains highest
levels in the pallidum and substantia nigra, with high levels
also in the striatum. Decreased GABA concentration has been
found in the caudate nucleus and putamen of brains from H.D.
patients (Perry, Hansen and Kloster, 1973). Further, glutamic
acid decarboxylase (GAD), the biosynthetic enzyme for GABA,
is decreased in the striatum and substantia nigra, but not in
the cortex of brains from H.D. patients (Bird and Iversen,
1974) (see Table 1). Though low GAD values have been noted in
brains obtained from patients likely to have suffered cerebral
hypoxia (Bowen, Smith, White and Davison, 1976; Perry, Gibson,
Blessed, Perry and Tomlinson, 1977), this selective loss does
not appear related to chronic hospitalization or neuroleptic
drug administration as normal GAD values are observed in long-
term psychiatric in-patients (Spokes, 1980). GABA concentrations appear fairly stable in post-mortem brain, and unlike GAD
are not influenced by the agonal state (Perry, Buchanan, Kish
and Hansen, 1979; Spokes, Garrett and Iversen, 1979).

We have focused our examinations on the substantia
nigra, which is often more darkly pigmented in H.D. than normal
(Bird, unpublished observations). The substantia nigra is
divided into two regions, the dorsal pars compacta and the
ventral pars reticulata. The pars compacta contains the cell
bodies of dopamine neurons whose axons form a pathway to the
striatum. Dendrites extend from the dopamine cell
body throughout both regions of the substantia nigra. The pars
reticulata receives axons from neuroinhibitory GABA cells in
the striatum, and the terminals of these axons are in contact
with the dendrites from dopamine cells. The dopamine concen-

TABLE 1

ACTIVITY OF GAD IN CONTROL AND
HUNTINGTON'S CHOREIC POST-MORTEM BRAIN

μ mol/$^{14}CO_2$ evolved/h/g tissue (mean $\pm$ S.E.M.)

AREA	CONTROL	HUNTINGTON'S CHOREA	P
GLOBUS PALLIDUS	7.1 ± 1.3 (24)	1.8 ± 0.5 (14)	0.001
SUBSTANTIA NIGRA	6.5 ± 1.0 (38)	2.2 ± 0.3 (40)	0.001
CAUDATE NUCLEUS	5.1 ± 0.4 (68)	1.5 ± 0.2 (66)	0.001
PUTAMEN	4.4 ± 0.4 (45)	0.9 ± 0.2 (41)	0.001
OLFACTORY TUBERCLE	3.1 ± 0.8 (11)	1.7 ± 0.3 (13)	N.S.
HYPOTHALAMUS	3.2 ± 0.5 (19)	4.4 ± 0.7 (18)	N.S.
FRONTAL CORTEX	3.0 ± 0.3 (30)	2.7 ± 0.3 (23)	N.S.
HIPPOCAMPUS	1.9 ± 0.4 (15)	2.0 ± 0.4 (12)	N.S.

tration in the pars compacta is normally twice that of the pars reticulata (Hornykiewicz, 1963), while the reverse is found for GABA concentration (Kanazawa, Miyata, Toyokura and Osuka, 1973). In H.D., the pars reticulata is much more atrophic than the pars compacta which is thought to be due to the loss of striatal afferents.

Comparisons of GABA concentrations in CSF between control and H.D. subjects have produced conflicting results. Manyam et al (1978) have reported a 50% reduction of GABA in CSF from H.D. subjects whereas other investigators have been unable to detect GABA in CSF from control subjects. Apparently, there are many difficulties with the measurement of GABA in CSF and a rigorous protocol for handling the CSF is required.

Paralleling the observation of decreased GABA is the increase in dopamine (DA) concentration found in the corpus striatum, and substantia nigra of H.D. patients (Spokes, 1980). Dopamine concentrations in choreic brain have been previously reported to be decreased in the caudate with no change in the putamen and substantia nigra (Bernheimer and Hornykiewicz, 1973). In the studies of Bird and Iversen (1974), the dopamine concentrations in caudate and putamen were not significantly different from controls in the putamen and caudate. However, after the method of freezing a brain as soon as removed from the skull was adopted, higher dopamine values have been obtained in control and choreic brain. The dopamine concentrations in all choreic basal ganglia have recently been noted to be greater than controls (Spokes, 1980). Using the method of high pressure liquid chromatography, Melamed, Hefti, Bird and Wurtman (unpublished) have found that there is a significant increase in dopamine concentration in the putamen but not in the caudate (see Table 2). These dopamine increases could not be attributed to chronic neuroleptic drug administration, as normal striatal dopamine values were found in samples from schizophrenic subjects treated with similar drug regimens (Bird, Spokes and Iversen, 1979). Moreover, drug-treated choreic subjects had dopamine values similar to their drug-free counterparts (Spokes, 1980).

Three possible factors may be affecting dopamine concentrations in H.D. brains. First, there may be functional increases due to the loss of neuroinhibitory influence of GABA in the substantia nigra. Second, there may be a relative increase due to loss of surrounding neurons. Finally, in those areas where there is marked atrophy but slightly increased concentrations of dopamine, there may be actual decreases due to loss of dopamine terminals. This may explain why there is less total dopamine in the caudate where the greatest loss of neurons occurs.

Since there is less striatal atrophy in the brain of patients that die in the early stages of chorea, there would no doubt be different concentrations of the various transmitters in such tissues. We have not had many cases to study that died early in their course. However, it has been our impression that many of the cases that have rigidity in the last few years of their life have more striatal atrophy. This would also agree with the previous reported findings in children where there is extreme atrophy of the caudate and putamen (Campbell, Corner, Norman and Urich, 1961).

We would propose that in the early stages of H.D. there may be an increased production and turnover of dopamine followed in time by a progressive decrease in the total amount of dopamine in the basal ganglia.

Dopamine receptor binding is reduced by 43% in the choreic caudate nucleus (Reisine, Fields, Bird, Spokes and Yamamura, 1978). This is a little less than the average total weight loss of the choreic caudate which is about 50%. Therefore, it is conceivable that the loss of post-synaptic neurons for dopamine transmission might lead to some atrophy of dopamine terminals and this may account for the fact that dopamine concentrations are not as high as expected in such atrophic tissue.

A useful method of estimating the functional activity of dopamine neurons is to measure brain concentrations of homovanillic acid (HVA), the major dopamine metabolite. Reduced CSF HVA concentrations have been found in H.D. suggesting that the turnover of the remaining dopamine in the corpus striatum is normal or slightly reduced (Curzon, Gumpert and Sharpe, 1972; Chase, 1973). However, HVA concentration in CSF may also depend on ventricular volume, and the decreased concentration in H.D. patients may be a reflection of ventricular enlargement (Curzon, 1975). CSF levels of HVA were found to be particularly low in juvenile patients with the rigid form of the disease (Curzon et al, 1972; Johansson and Roos, 1975) which is characterized by marked atrophy of the corpus striatum.

Relatively few studies on H.D. brain have reported on HVA measurements. Bernheimer and Hornykiewicz (1973) noted that HVA was not significantly different in the putamen of H.D. brain, and was decreased in the caudate nucleus (see Table 2). Melamed, Hefti, Bird and Wurtman (unpublished) have recently measured HVA concentration in choreic brain and found no significant differences (see Table 2). Dopamine measurements on the same samples revealed significant increased in dopamine concentrations in H.D. putamen, but without an increase in the

206

TABLE 2

BRAIN DOPAMINE AND HOMOVANILLIC ACID IN HUNTINGTON'S DISEASE

	DOPAMINE µg/g tissue		HVA µg/g tissue	
	CAUDATE	PUTAMEN	CAUDATE	PUTAMEN
BERNHEIMER & HORNYKIEWICZ ET AL (1973)				
CONTROLS	2.6 ± 0.3 (28)	3.4 ± 0.3 (28)	3.2 ± 0.3 (8)	4.3 ± 0.7 (8)
H.D.	1.6 ± 0.2 (10)	2.7 ± 0.4 (10)	2.2 ± 0.5 (8)	3.9 ± 0.6 (8)
MELAMED ET AL (1980)				
CONTROLS	3.2 ± 0.7 (11)	3.0 ± 0.6 (14)	1.9 ± 0.2 (11)	2.7 ± 0.3 (14)
H.D.	3.5 ± 0.7 (11)	6.0 ± 1.2 (11)*	1.9 ± 0.3 (11)	2.9 ± 0.4 (11)

* P <0.05 AS COMPARED TO CONTROLS (T-TEST)

VALUES EXPRESSED AS ± SEM. NUMBER OF SAMPLES IN PARENTHESIS

caudate nucleus (see Table 2). There appears to be increased evidence of a difference between the dopamine concentration in the caudate and putamen in chorea with the putamen generally exhibiting higher dopamine concentrations. Whether this might be due to a difference in the rate of atrophy is unknown at this time.

Acetylcholine has long been considered an antagonist of dopamine in the basal ganglia. The enzyme choline acetyl-transferase (CAT), responsible for the biosynthesis of acetylcholine, is considered to be exclusively localized in cholinergic neurons. It is highly active in basal ganglia, especially putamen. Although there is an average loss of CAT activity of over 50% per unit weight in putamen and caudate nucleus (see Table 3), a substantial portion of choreic cases show CAT activity within the normal range in basal ganglia (Bird and Iversen, 1974). The remaining cases show approximately 85% reduction of CAT activity in basal ganglia. CAT activity is only slightly reduced in the nucleus accumbens of H.D. patients (Spokes, 1980). This is consistent with other reports detailing the relative sparing of this region compared with striatum (Aquilonius, Eckernas and Sundwall, 1975; McGeer and McGeer, 1976). Normal CAT values were obtained in both striatum and other brain regions from long-term psychiatric inpatients, indicating that neither hospitalization nor neuroleptic drug treatment is responsible for these changes in H.D. (Bird et al, 1979).

Serotonin-containing nerve fibers arise from cell bodies in the mid-brain raphe nuclei and end in a variety of brain regions. Serotonin concentrations in various brain regions from six H.D. cases were examined by Bernheimer and Hornykiewicz (1973), and although there were increased concentrations in the putamen, globus pallidus and central gray of the midbrain, these were not considered to be significantly different from normal. However, Curzon, in a larger unpublished analysis, found significant increases in serotonin in the caudate of H.D. brains, and the levels were directly proportional to the decrease in GAD activity. CSF values of 5-hydroxyindoleacetic acid, a serotonin metabolite, are reportedly within the normal range in choreic subjects, both at steady state and after probenecid loading (Chase, 1973).

There does not appear to be any loss of serotonergic neurons in H.D.

A number of neuropeptides in human brain are now being measured (see Table 4). Angiotensin-converting enzyme (ACE) converts the inactive decapeptide angiotensin I to the active octapeptide angiotensin II, and is ubiquitous in brain tissue

TABLE 3

ACTIVITY OF CHAc IN CONTROL AND
HUNTINGTON'S CHOREIC POST-MORTEM BRAIN

Enzyme activity (μmol/h/g tissue) mean $\pm$ S.E.M.

	CONTROL	HUNTINGTON'S CHOREA	P
PUTAMEN	$21.8 + 1.6$ (41)	$9.1 + 1.3$ (42)	0.001
CAUDATE NUCLEUS	$11.9 + 0.8$ (60)	$5.4 + 0.7$ (64)	0.001
OLFACTORY TUBERCLE	$2.9 + 0.4$ (15)	$2.5 + 0.7$ (15)	N.S.
SUBSTANTIA NIGRA	$1.2 + 0.4$ (12)	$1.0 + 0.5$ (14)	N.S.
FRONTAL CORTEX	$1.2 + 0.1$ (14)	$1.4 + 0.1$ (11)	N.S.
GLOBUS PALLIDUS	$0.9 + 0.2$ (9)	$1.5 + 0.3$ (8)	N.S.

209

with high activities found in the corpus striatum. Angiotensin II may function as a neurotransmitter in brain (Barker, 1976). Selective depletion of ACE activity has been demonstrated in the corpus striatum of H.D. brains (Arregui, Iversen, Spokes and Emson, 1979). The decrease was 83 to 92% in globus pallidus; 62 to 69% in caudate and putamen. Normal values were obtained for the two cortical areas.

The neuropeptide, gonadotropin-releasing hormone (GRH), has been measured in the human hypothalamus (Bird, Chiappa and Fink, 1976; Okon and Koch, 1976) (see Table 4). Evidence suggests that GRH is modulated by dopaminergic neurons. GRH may thus be increased as is dopamine in H.D. In fact, increased GRH was observed in the median eminence of the female choreic brain, but not in the male (Bird et al, 1976). Increased fertility has been noted in female patients with H.D. Whether this is due to alterations in GRH remains to be determined.

Similarly, there is evidence that the concentration of substance-P in the brains of choreics is different from that in controls. Substance-P is an undecapeptide that is unevenly distributed in brain with the highest concentrations reported in the substantia nigra (Brownstein, Mroz, Tappaz and Leeman, 1977). Markedly decreased concentrations of substance-P have been found in the substantia nigra of choreic patients (Kanazawa, Bird, O'Connell and Powell, 1977). A decrease was also found in the globus pallidus, while normal levels were reported in cerebral cortex, caudate and putamen (Gale, Bird, Spokes, Iversen and Jessell, 1978)(see Table 5).

Substance-P terminals in the substantia nigra have been seen in close association with the dopamine dendrites (Ljungdahl, Hokfelt, Nilsson and Goldstein, 1978), and it has been thought that substance-P is modulating the inhibitory stimulus of GABA on the dopaminergic cell. Whether the loss of substance-P in the SN is due to degeneration of substance-P terminals is less clear. Studies in animals with lesions placed between the caudate and the SN caused a decrease in substance-P in the SN suggesting a striatial-nigral pathway (Kanazawa, Emson and Cuello, 1977). If the loss of substance-P in H.D. were due to degeneration of substance-P cell bodies in the caudate then we would expect to find a decrease in substance-P concentration in the caudate, and this is not the case (Gale et al, 1978).

It is possible, therefore, that the loss of GABA in the SN may precede the loss of substance-P and result in a feedback reduction of substance-P in the SN. The substance-P cell

TABLE 4

NEUROPEPTIDES IN HUMAN BRAIN

	Hypothalamus	Amygdala	Substantia Nigra	Cortex	Reference[*]
GRH[1]	27	--	--	<4	1
TRH[2]	54	--	--	1	2
Substance-P[3]	5	3	47	2	3
Neurotensin[4]	32	5	23	1	4
Somatostatin[4]	278	339	24	53	4
Gastrin/CCK[5]	6	--	--	199	5
VIP[6]	23	21	2	17	6

1 = pg /mg^{-1}/tissue
2 = pg /mg /tissue
2 = pmol/mg /protein
4 = pg / g /tissue
5 = pmol/ g /tissue

[*]REFERENCES:

1. Bird et al (1976)
2. Okon and Koch (1976)
3. Gale et al (1978)
4. Cooper et al (1980)
5. Vanderhaeghen et al (1975)
6. Emson et al (1979)

211

TABLE 5
HUMAN BRAIN PEPTIDES IN HUNTINGTON'S DISEASE VS CONTROLS

	CONTROL	HUNTINGTON'S DISEASE	REFERENCE
GRH[1]			
MEDIAN EMINENCE	314 ± 84 (11)	1231 ± 410 (9)*	1
SOMATOSTATIN[2]			
HYPOTHALAMUS	1.9 ± 0.2 (19)	3.9 ± 0.4 (20)**	2
SUBSTANCE-P[3]			3
SUBSTANTIA NIGRA	47.2 ± 4.8 (13)	22.2 +2.3 (9)**	
GLOBUS PALLIDUS	18.0 ± 3.3 (18)	9.7 ± 1.9 (15)**	
CAUDATE	3.7 ± 0.8 (18)	3.5 ± 0.6 (19)	
FRONTAL CORTEX	2.4 ± 0.3 (9)	1.7 ± 0.3 (7)	
VIP[4]			4
FRONTAL CORTEX	17.3 ± 2.3 (21)	14.2 ± 1.5 (11)	
CAUDATE	4.6 ± 1.3 (12)	4.2 ± 2 (12)	
CCK-8[4]			5
SUBSTANTIA NIGRA	65 ± 10 (10)	25 ± 3 (10)**	
GLOBUS PALLIDUS	21 ± 4 (10)	10 ± 2 (10)**	
CAUDATE	79 ± 23 (10)	71 ± 6 (10)	
FRONTAL CORTEX	137 ± 14 (10)	138 ± 16 (10)	

VALUES EXPRESSED AS ± SEM FOR NUMBER OF SAMPLES IN PARENTHESIS
TWO TAILED T TEST * P 0.05 ** P 0.001

1=PG /MG^{-1}/TISSUE
2=NG /MG /PROTEIN
3=PMOL/MG /PROTEIN
4=PMOL/ G /TISSUE

REFERENCES:
1. BIRD ET AL 1976
2. COOPER ET AL 1980
3. GALE ET AL 1978
4. EMSON ET AL 1979
5. EMSON ET AL 1980

bodies in the caudate may remain intact because of collaterals
to other regions that are not atrophic in H.D.

There is also another possible mechanism for the maintenance of substance-P concentrations in the choreic caudate.
Recent immunocyto-chemical studies have shown that substance-P
probably exists within serotonin neurons (Hokfelt, Ljungdahl,
Steinbusch, Verhofstad, Nilsson, Brodin, Pernow and Goldstein,
1978). Serotonergic neurons do not appear to degenerate in H.D.
(Curzon-unpublished), so it is possible that the substance-P
within serotonergic neurons in the striatum is also maintained
in H.D. There are very few serotonergic neurons in the SN.
This indicates the importance of combining immunocytochemistry
with quantitative measurements of neuropeptides. This has not
been done as yet in H.D. brain tissue.

Vanderhaeghen, Signeau and Gepts (1975) reported a
peptide present in human brain that reacted with antigastrin
antibodies. This brain gastrin peptide (BGP) was found in much
higher concentrations in cerebral cortex than in basal ganglia
and brain stem. Subsequent studies (Robberecht, Deschodt-
Lanckman and Vanderhaeghen, 1978; Rehfeld, 1978) indicated that
the antigastric antibody was also measuring a similar four peptide sequence that appears at the COOH-terminal of cholecystokinin (CCK). With the recent production of specific antibodies
to other portions of the CCK peptide (Rehfeld, 1980), it would
appear that most of what was being measured earlier as gastrin
in the cortex is CCK (Table 4). The biological activity of CCK
lies within the terminal eight peptides.

Since this peptide appears to affect satiety (Gibbs,
Young and Smith, 1973), it is conceivable that the continual
hunger that so many choreics exhibit may be due to alterations
of this peptide in the H.D. brain.

Measurements of CCK-8 in H.D. brain reveal that the
only areas that show significant differences are decreased concentrations in the globus pallidus and SN (Emson et al, 1980).
There are no significant differences found in the cortex, or
caudate and putamen of H.D. brain. CCK-8 is similar to substance-P in this selective loss in H.D., since substance-P
is also decreased in globus pallidus and SN, but not decreased
in cerebral cortex, or caudate and putamen (Gale et al, 1978).
(See Table 5.)
Both of these peptides have been associated with dopaminergic neurons, CCK-8 being found within the terminals
(Hokfelt, 1980), and substance-P found on dopamine dendrites in
SN (Ljungdahl et al, 1978). In H.D., the activity of the

213

dopaminergic neurons in the striatum are maintained, and therefore this is the likely explanation for the maintenance of
CCK-8 concentrations in H.D. striatum.

Vasoactive Intestinal Polypeptide (VIP) is also one of
the intestinal peptides that is found in high concentrations in
cerebral cortex. This peptide was measured in the cerebral cortex and caudate of control and choreic brain, and no differences
were noted (Emson, Fahrenkrug and Spokes, 1979)(see Table 5).

CONCLUSION

The neurochemical studies described above have given us
some clues regarding the type of neurons affected in this
tragic disorder. However, we still do not know what initiates
cell death. The availability of new techniques such as immunocytochemistry will hopefully, in the future, help us in localizing early cell changes in choreic cases that may die prema-
turely.

CREDITS

TABLE 1

Data from Bird (1976)

TABLE 2

Data from Bernheimer & Hornykiewicz et al (1973) -- see reference list; Melamed et al (1980) -- unpublished data

TABLE 3

Data from Bird and Iversen (1974)

TABLE 4

References on Table. Cooper et al (1980) - unpublished data

TABLE 5

References on Table. Cooper et al (1980) - unpublished data

REFERENCES

Alzheimer, A. (1911). Huntingtoniche chorea und die chorea-
tishen Bewegungen iberthaupt. Z. Gesamte Neurol.
Psychiatr. 3:891-892.

Aquilonius, S.M., Eckernas, S.A. and Sundwall, A. (1975).
Regional distribution of choline acetyltransferase in human
brain: changes in Huntington's chorea. J. Neurol. Neurosurg.
Psychiatry 38:669-677.

Arregui, A., Iversen, L.L., Spokes, E.G.S. and Emson, P.C.
(1979). Alterations in postmortem brain angiotensin-
converting enzyme activity and some neuropeptides in Huntington's disease. In: Advances in Neurology (eds. T.N.
Chase, N.S. Wexler and A. Barbeau), Raven Press, New York,
23:517-525.

Barker, J.L. (1976). Peptides: roles in neuronal excitability.
Physiol. Rev. 56:435-452.

Bernheimer, H. and Hornykiewicz, O. (1973). Brain amines in
Huntington's chorea. In: Advances in Neurology (eds. A.
Barbeau, T.N. Chase and G.W. Paulson), Raven Press, New York,
1:525-531.

Bird, E.D. (1976). Biochemical studies on γ-aminobutyric acid
metabolism in Huntington's chorea. In: Biochemistry and
Neurology (eds. H.F. Bradford and C.D. Marsden), Academic
Press, London, pp. 83-91.

Bird, E.D. and Iversen, L.L. (1974). Huntington's chorea:
Post-mortem measurement of glutamic acid decarboxylase,
choline acetyltransference and dopamine in basal ganglia.
Brain 97:457-472.

Bird, E.D., Chiappa, S.A. and Fink, G. (1976). Brain immuno-
reactive gonadotripin-releasing hormone in Huntington's
chorea and in non-choreic subjects. Nature 260:536-538.

Bird, E.D., Spokes, E.G.S. and Iversen L.L. (1979). Increased
dopamine concentration in limbic areas of brain from patients
dying with schizophrenia. Brain 102:347-360.

Bowen, D.M., Smith, C.B., White, P. and Davison, A.N. (1976).
Neurotransmitter-related enzymes and indices of hypoxia in
senile dementia and other abiotrophies. Brain 99:459-496.

Brownstein, M.J., Mroz, E.A., Tappaz, M.L. and Leeman, S.E. (1977). On the origin of substance P and glutamic acid decarboxylase (GAD) in the substantia nigra. Brain Research 135:315-323.

Bruyn, G.W. (1968). Diseases of the basal ganglia. In: Handbook of Clinical Neurology, (eds. P.J. Vinken and G.W. Bruyn), North-Holland Publishing Company, Amsterdam, 6:298-378.

Campbell, A.M.G., Corner, B., Norman, R.M. and Urich, H. (1961). A rigid form of Huntington's disease. J. Neurol. Neurosurg. Psychiatry 24:71-77.

Chase, T.N. (1973). Biochemical and pharmacological studies of monoamines in Huntington's chorea. In: Advances in Neurology, (eds. A. Barbeau, T.N. Chase and G.W. Paulson), Raven Press, New York, 1:533-542.

Curzon, G., Gumpert, J. and Sharpe, D. (1972). Amine metabolites in the cerebral spinal fluid in Huntington's chorea. J. Neurol. Neurosurg. Psychiatry 35:514-519.

Curzon, G. (1975). CSF homovanillic acid: an index of dopaminergic activity. In: Advances in Neurology, (eds. D. Calne, T.N. Chase and A. Barbeau), Raven Press, New York, 9:349-357.

Emson, P.C., Fahrenkrug, J. and Spokes, E.G.S. (1979). Vasoactive intestinal polypeptide (VIP): distribution in normal human brain and in Huntington's disease. Brain Research, 173:174-178.

Emson, P. et al (1980). Abstract - Transmitter Biochemistry of Human Brain Tissue Symposium, Goteborg, Sweden.

Folstein, S.E., Folstein, M.F. and McHugh, P.R. (1979). Psychiatric syndromes in Huntington's disease. In: Advances in Neurology, (eds. T.N. Chase, N.S. Wexler and A. Barbeau), Raven Press, New York, 23:281-289.

Gale, J.S., Bird, E.D. Spokes, E.G.S., Iversen, L.L. and Jessell, T. (1978). Human brain substance P: distribution in controls and Huntington's chorea. J. Neurochem. 30:633-634.

Gibbs, J., Young, R.C. and Smith, G.P. (1973). Cholecystokinin elicits satiety in rats with open gastric fistulas. Nature 245:328-330.

Hokfelt, T., Ljungdahl, A., Steinbusch, H., Verhofstad, A.,
Nilsson, G., Brodin, E., Pernow, B. and Goldstein, M. (1978).
Immunohistochemical evidence of substance-P-like immunoreactivity in some 5-hydroxytryptamine-containing neurons in
the rat central nervous system. Neuroscience 3:517-538.

Hokfelt, T. et al (1980). Eur. J. Pharmacol. (submitted).

Hornykiewicz, O. (1963). Die topische Lokalisation und das
Verhalten von Noradrenalin und Dopamin (3-hydroxytyramin) in
der Stubstantia nigra des normalen und Parkinson-kranken
Menschen. Wien. Klin. Wschr. 75:309-312.

Johansson, B. and Roos, B.E. (1975). Concentrations of monoamine metabolites in human lumbar and cisternal cerebrospinal
fluid. Acta Neurol. Scand. 52:137-144.

Kanazawa, I., Miyata, Y., Toyokura, Y. and Otsuka, M. (1973).
The distribution of γ-aminobutyric acid (GABA) in the human
substantia nigra. Brain Res. 51:363-365.

Kanazawa, E., Bird, E.D., O'Connell, R. and Powell, D. (1977).
Evidence for a decrease in substance P content of substantia
nigra in Huntington's chorea. Brain Res. 120:387-392.

Kanazawa, I., Emson, P.C. and Cuello, A.C. (1977). Evidence
for the existence of substance P-containing fibres in
striato-nigral and pallido-nigral pathways in rat brain.
Brain Res. 119:447-453.

Lange, H., Thorner, G., Hopf, A. and Schroder, K.F. (1976).
Morphometric studies of the Neuro-pathological changes in
choreatic diseases. J. Neurol. Sci. 28:401-425.

Lieberman, A., Dziatolowski, M., Neophytides, A., Kupersmith,
M., Aleksic, S., Serby, M., Korein, J. and Goldstein, M.
(1979). Dementias of Huntington's and Parkinson's disease.
In: Advances in Neurology, (eds. T.N. Chase, N.S. Wexler,
and A. Barbeau), Raven Press, New York, 23:273-280.

Ljungdahl, A., Hokfelt, T., Nilsson, G. and Goldstein, M.
(1978). Distribution of substance P-like immunoreactivity
in the central nervous system of the rat-II. Light microscopic localization in relation to catecholamine-containing
neurons. Neuroscience 3:945-976.

Manyam, N.J., Hare, T.A., Katz, L. and Glaeser, B.S. (1978). Huntington's disease. Cerebrospinal fluid GABA levels in at-risk individuals. Arch. Neurol. $\underline{35}$:728-730.

McGeer, P.L. and McGeer, E.G. (1976). Enzymes associated with the metabolism of catecholamines, acetylcholine and GABA in human controls and patients with Parkinson's disease and Huntington's chorea. J. Neurochem. $\underline{26}$:65-76.

Okon, E. and Koch, Y. (1976). Nature $\underline{263}$:345-347.

Oliver, J.E. (1970). Huntington's chorea in Northamptonshire. Br. J. Psychiatry $\underline{116}$:241.

Perry, E.K., Gibson, P.N., Blessed, G., Perry, R.H. and Tomlinson, B.E. (1977). Neurotransmitter enzyme abnormalities in senile dementia. J. Neurol. Sci. $\underline{34}$:247-265.

Perry, T.L., Hansen, S. and Kloster, M. (1973). Huntington's chorea: deficiency of γ-aminobutyric acid in brain. N. Eng. J. Med. $\underline{288}$:337-342.

Perry, T.L., Buchanan, J., Kish, S.J. and Hansen, S. (1979). γ-aminobutyric acid deficiency in brain of schizophrenic patients. Lancet $\underline{i}$:237-239.

Reed, T.E. and Neel, J.V. (1959). Huntington's chorea in Michigan. II. Selection and mutation. Amer. J. Hum. Genet. $\underline{II}$:107-136.

Rehfeld, J.F. (1978). Immunochemical studies on cholecystokinin. J. Biol. Chem. $\underline{253}$:11, 4022-4030.

Rehfeld, J.F. (1980). Cholecystokinin. Trends in Neuro-Sciences $\underline{3}$:65-67.

Reisine, T.D., Fields, J.Z., Bird, E.D., Spokes, E. and Yamamura, H.I. (1978). Characterization of brain dopaminergic receptors in Huntington's disease. Commun. Psychopharmac. $\underline{2}$:79-84.

Robberecht, P., Deschodt-Lanckman, M. and Vanderhaeghen, J.J. (1978). Demonstration of biological activity of brain gastrin-like peptidic material in the human: Its relationship with the COOH-terminal octa-peptide of cholecystokinin. Proc. Natl. Acad. Sci. USA $\underline{75}$:1, 524-528.

Spokes, E.G.S., Garrett, N.J. and Iversen, L.L. (1979).
Differential effects of agonal status on measurements of
GABA and glutamate decarboxylase in human post-mortem
brain tissue from control and Huntington's chorea subjects.
J. Neurochem. 33:773-778.

Spokes, E.G.S. (1980). Neurochemical alterations in Huntington's chorea - A study of post-mortem brain tissue.
Brain 103:179-210.

Tellez-Nagel, I., Johnson, A.B. and Terry, R.D. (1973).
Ultrastructural and histochemical study of cerebral biopsies in Huntington's chorea. In: Advances in Neurology
(eds. A. Barbeau, T.N. Chase and G.W. Paulson), Raven Press,
New York, 1:387-398.

Vanderhaeghen, J.J., Signeau, J.C. and Gepts, W. (1975). New
peptide in the vertebrate CNS reacting with antigastrin
antibodies. Nature 257:604-605.

Wallace, D.C. (1974). Huntington's chorea in Queensland. A
not uncommon disease. Med. J. Aust. 39:375.

Wendt, G.G. and Drohm, D. (1972). (Huntington's chorea. A
population-genetic study). In: Humangenetik: Advances in
Human Genetics, Theime, Stuttgart, 4:1-121.

Westphal, A. (1905). Case Report. Zbl. Nervenheilk 28:674-675.

Neuropeptides in Human Brain: Studies on Substance P, Cholecystokinin, Vasoactive Intestinal Polypeptide and Methionine-Enkephalin in Normal Human Brain and in Huntington Disease

P. C. EMSON*, M. ROSSOR*, S. P. HUNT*, V. CLEMENT-JONESt,
J. FAHRENKRUG‡ and J. REHFELD¶

* MRC Neurochemical Pharmacology Unit, Dept. of
  Pharmacology, Medical School, Hills Road,Cambridge, England
† Dept. of Chemical Pathology, St Bartholomew's Hospital, London
‡ Dept. of Clinical Chemistry, Bispeberg Hospital, Copenhagen, Denmark
¶ Dept. of Medical Biochemistry, University of Aarhus, Denmark

In recent years it has become clear that there are a number of neuronally localized peptides present in the mammalian central nervous system. Current evidence on the physiological roles of many of these peptides is very incomplete but some at least may be considered as candidates for neurotransmitters or neurohormones (see reviews by Emson 1979, Snyder 1980). The post-mortem human brain contains significant amounts of vasoactive intestinal polypeptide (VIP) (Emson et al. 1979a), substance P, (Kanazawa et al. 1977, Gale et al. 1978) cholecystokinin octapeptide (Emson et al. 1979b, Emson et al. 1980b) and several endogenous opioid peptides including methionine-enkephalin and β-endorphin (Gramsch et al. 1979, 1980, Emson et al. 1980a).

Materials and Methods

The collection, storage and dissection of human brains post-mortem was carried out as described previously (Spokes, 1980). In model experiments to study the stability of peptides in animal brain post-mortem the approach described by Spokes and Koch (1978) was used, and mice were killed and cooled gradually in a programmed incubator to simulate the cooling curve determined previously for human brains post-mortem (Spokes and Koch, 1978). Mouse brains were removed at different times during a 72 hour period (see Figure 1) and their peptide contents determined by radioimmunoassay.

Mouse or human brain samples were weighed and extracted using either boiling water followed by subsequent acetic acid extraction (for cholecystokinin-like peptides) (Rehfeld, 1978) or boiling 1M acetic acid (for substance P, met-enkephalin and VIP).

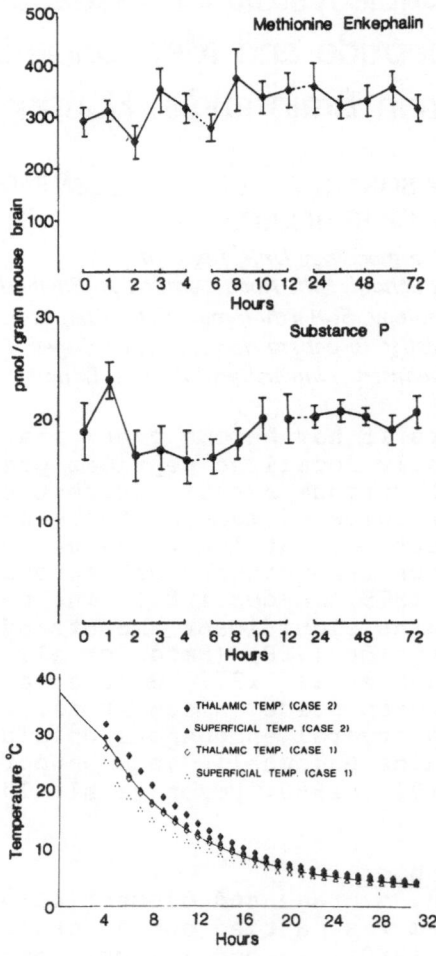

Figure 1. Post-mortem stability of methionine-enkephalin and substance P-like immunoreactivity in mouse brains. Mice were killed and kept in an incubator for up to 72 hours. The incubator was programmed to cool the mouse brain from 37°C to 4°C over a period of 32 hours matching the cooling curve determined for human brain (Spokes and Koch 1978). (a) methionine enkephalin, (b) substance P and (c) human brain cooling curves from Spokes and Koch (1978).

222

The supernatants from these extractions were freeze-dried and assayed as described (Rehfeld, 1978, Kanazawa and Jessell 1976, Emson et al. 1980a and Fahrenkrug and Muckadell 1977). Chromatographic separations of cholecystokinin and VIP-like peptides in human brain were carried out on Sephadex G-50 columns, enkephalin and substance P-like immunoreactivities were separated on Bio-Gel P2 columns and further purified by high pressure liquid chromatography (Emson et al. 1980a, Morris et al. 1980).

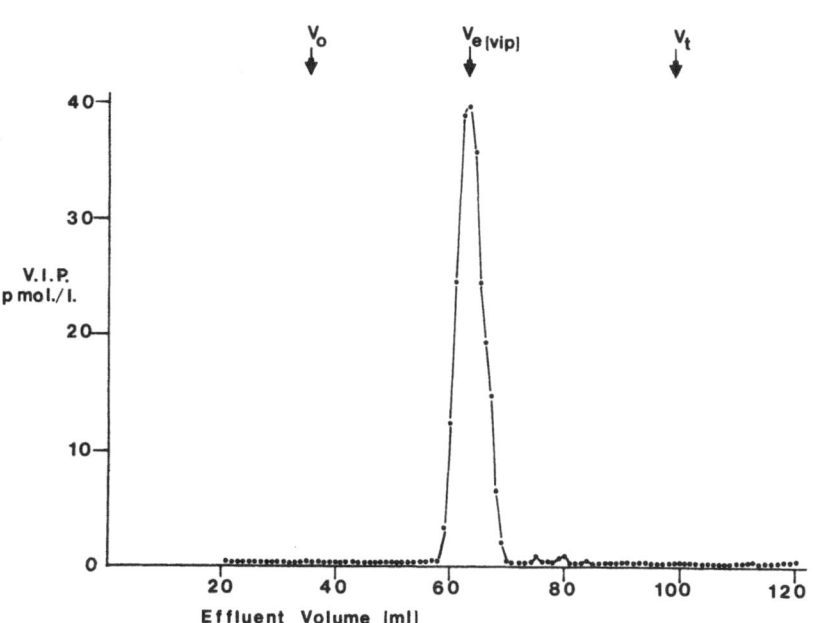

Figure 2. Elution pattern of VIP-like immunoreactivity from human brain cerebral cortex (Sephadex G-50 column). Note the presence of a VIP-like peptide which elutes from the column at the position (Ve) expected for the intact twenty-eight amino acid porcine VIP.

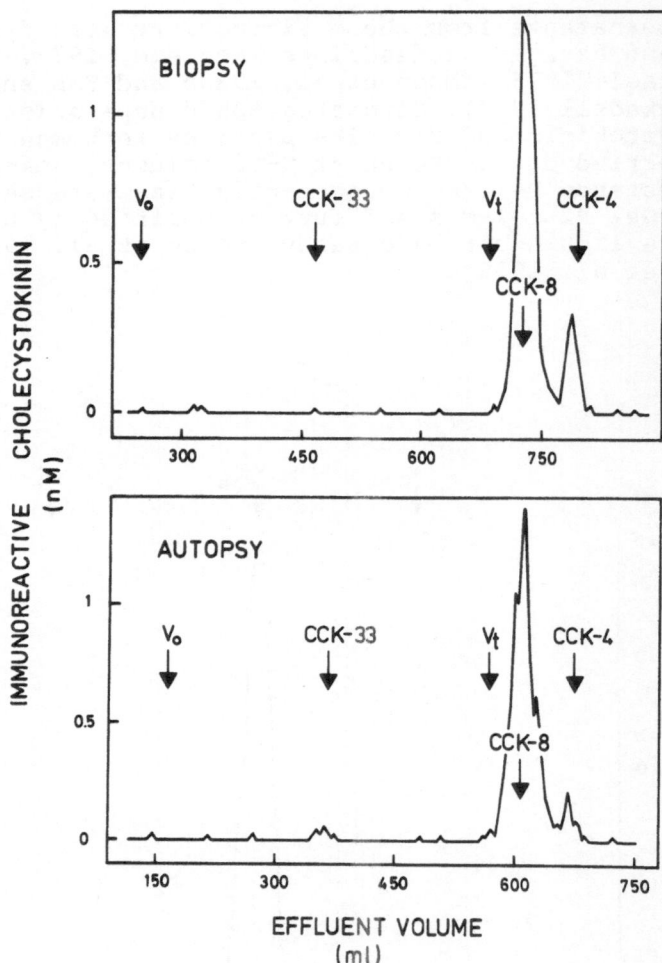

Figure 3.   Comparison of the chromatographic behaviour
of cholecystokinin-like immunoreactivity extracted
from autopsy or biopsy material of human cerebral
cortex. Note the presence of the predominant molecular
form the octapeptide peptide (CCK-8). Sephadex G-50
column.

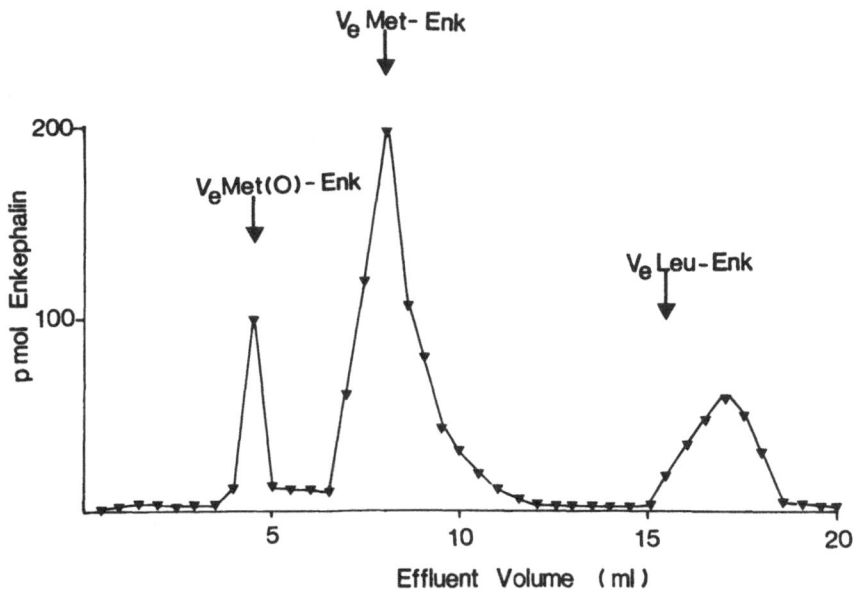

Figure 4. High pressure liquid chromatography of
human brain globus pallidus extract (lateral segment).
Note the presence of separate methionine and leucine
enkephalin peaks in the extracts. There is also a
small amount of oxidized methionine enkephalin. HPLC
carried out on a μ-Bondapak C18 column as described by
Morris et al. (1980).

Post-mortem Stability of Substance P, Cholecystokinin, VIP and met-enkephalin in Human Brain

Detailed study of the peptide content of mouse
brains cooled in a programmed incubator showed that
none of the peptides studied showed significant post-
mortem degradation (Figure 1). That this is also
likely to be the case in the post-mortem human brain
has been shown by a series of studies which show that
the neuropeptides detected by radioimmunoassay in the
human post-mortem brain correspond to the intact
peptides and not to fragments (see Figures 2,3 and 4
and also Emson et al. 1980a), and further comparison
of the content of the cortical peptides (VIP/CCK) in
human biopsy material revealed the presence of amounts
of these peptides similar to those found in human
autopsy material (Figure 3). All this data suggests

strongly that as in the mouse brain, the neuropeptides measured in this study may be equally stable in human brain tissue post-mortem. This suggestion is further supported by our observations that immunohistochemical techniques will demonstrate concentrations of enkephalin and substance P-like immunoreactivity in human brain areas which are known to contain enkephalin and substance P neurones and terminals in the rat brain. Thus, in Figure 5 the substance P immunoreactivity is concentrated in the pars reticulata of the human substantia nigra and there is no suggestion of extensive diffusion of substance P away from its location.

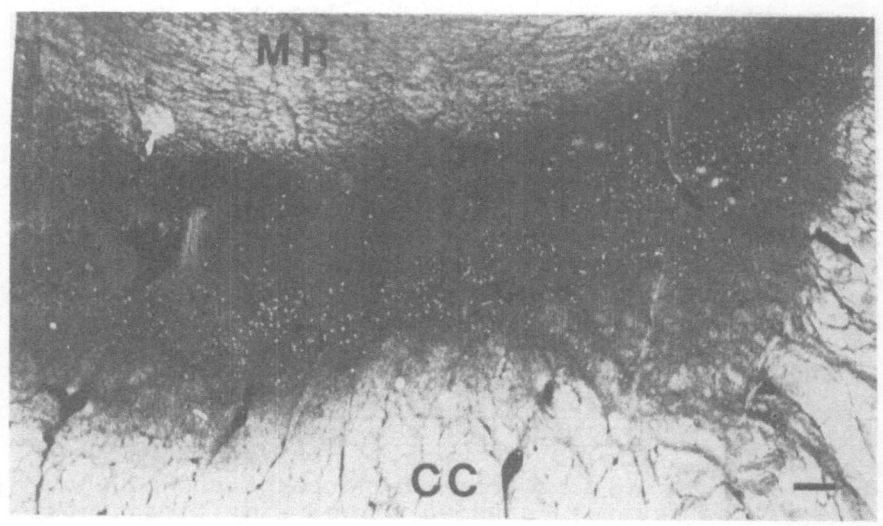

Figure 5. Substance P immunoreactivity within the human substantia nigra (SN). Substance P immunoreactivity is largely found in areas devoid of pars compacta neurones. The compacta neurones are revealed through their endogenous peroxidase activity (◁). MR, Mesencephalic reticular formation, CC, crus cerebri, Scale bar = 400µm.

TABLE 1. The regional distribution of cholecystokinin (CCK-8 equivalents) and vasoactive intestinal polypeptide in normal human brain

	CCK (pmol/g)	VIP (pmol/g)		CCK (pmol/g)	VIP (pmol/g)
Cerebral cortex			Caudate	79 ± 23	4.6 ± 1.3
Brodman area 4	159 ± 12	7.9 ± 1.4	Putamen	67 ± 6	2.3 ± 0.5
Brodman area 6	93.4 ± 6	15.7 ± 2.0	Globus pallidus (medial)	21 ± 4	2.4 ± 1.4
Brodman area 8	60.1	10.2 ± 1.4			
Brodman area 10	137 ± 45	17.3 ± 2.3			
Brodman area 11	108	22.8 ± 2.6	Hypothalamus	N.D	23.1 ± 6.0
Brodman area 32	197	21.4 ± 3.1	Substantia nigra (pars compacta/pars reticulata)	65 ± 10	1.9 ± 1.3
Brodman area 38	76.1 ± 5.3	10.9 ± 0.4	Anteromedial thalamus	3 ± 1	0.3 ± 0.08
Brodman area 31	97	16.7 ± 2.8	Central grey	11.7	2.8 ± 0.5
Hippocampus	109.6	10.7 ± 1.0	Cerebellum	1.0	0.4 ± 0.1
Amygdala	96.5	20.8 ± 8.9	Olive	2.53	–

Values are means ± S.E.M. for at least 6 separate determinations, where less than this number of determinations have been carried out a mean value only is given.

TABLE 2. The regional distribution of methionine enkephalin and substance P in normal human brain

	Met-Enk (pmol/g)	Substance P (pmol/g)
Cerebral cortex (area 10)	42 ± 17	14 ± 10
Caudate nucleus	116 ± 20	138 ± 14
Putamen	210 ± 71	112 ± 29
Globus pallidus (lateral)	1163 ± 216	197 ± 99
Globus pallidus (medial)	675 ± 168	877 ± 253
Nucleus accumbens	1086 ± 240	159 ± 21
Amygdala	26 ± 10	25 ± 12
Hippocampus	56 ± 16	104 ± 29
Hypothalamus	141 ± 22	112 ± 19

	Met-Enk (pmol/g)	Substance P (pmol/g)
Substantia nigra (pars compacta)	557 ± 163	1264 ± 239
Substantia nigra (pars reticulata)	661 ± 145	1535 ± 177
Interpeduncular nucleus	–	83 ± 25
Superior colliculus	55 ± 4	–
Central grey	143 ± 46	378 ± 130
Locus coeruleus	79 ± 18	310 ± 72
Raphe Pons	70 ± 25	234 ± 23
Basal Pons	10 ± 2	5 ± 1
Pineal gland	–	N.D

Values are means ± S.E.M. for between 5-40 separate normal brains.

Regional Distribution in Normal Human Brain

The regional distributions of cholecystokinin and vasoactive intestinal polypeptide in normal human brain are summarized in Table 1. Both VIP and CCK-like peptides are concentrated particularly in forebrain areas including hypothalamus, basal ganglia and cerebral cortex. The distribution of both peptides in human brain agrees well with the regional distributions found in other mammals (Larsson and Rehfeld 1979, Fahrenkrug 1979) and the contents of both peptides (VIP and CCK) seems to correlate with the neuronal density of human brain. In the rat brain both CCK and VIP positive neurones can be demonstrated by immuno-histochemical technique and in addition ascending CCK and VIP containing projections running into the forebrain can be demonstrated (Marley, Emson and Hunt unpublished observations).

The regional distributions of methionine-enkephalin and substance P immunoreactivity in human brain are summarized in Table 2. Again as for VIP and CCK, the distributions of both methionine-enkephalin and substance P agree well with previous studies on laboratory mammals (Kanazawa and Jessell 1976, Hong et al. 1977, Wesche et al. 1977). The highest concentrations of both substance P and methionine-enkephalin are found in the human basal ganglia with methionine-enkephalin being particularly concentrated in the globus pallidus (lateral segment) and nucleus accumbens (in very good agreement with histochemical observations, Cuello 1977) and substance P concentrated particularly in the substantia nigra and globus pallidus (medial segment) (see also Figure 5). The majority of enkephalin-like immunoreactivity in the human globus pallidus corresponds to methionine-enkephalin (Figure 4) with a ratio of methionine : leucine enkephalin of between 4 and 5:1 in this area. In the substantia nigra there was also a high molecular weight endorphin-like peptide and the content of this material is not changed in Huntington's disease although the amount of authentic met-enkephalin is significantly reduced (Table 3).

Huntington's Disease

Because of the reported pathology of the basal ganglia and cerebral cortex in Huntington's disease it was of considerable interest to examine the content of substance P, methionine-enkephalin, vasoactive intestinal polypeptide and cholecystokinin in these areas in the brains of patients dying with Huntington's disease. The data summarized in Table 3, (compare with

TABLE 3. Content of substance P, methione-enkephalin, cholecystokinin and vasoactive intestinal peptide in brain regions in Huntington's disease

	Substance P (pmol/g)	Met-Enk (pmol/g)	CCK (pmol/g)	VIP (pmol/g)
Caudate	158 ± 28 (10)	156 ± 54 (10)	71 ± 6 (10)	4.2 ± 2.0 (12)
Putamen	77 ± 20 (10)	184 ± 68 (12)	77 ± 13 (10)	2.5 ± 2.0 (20)
Globus pallidus (lateral)	18 ± 9* (10)	527 ± 102* (39)	-	-
Globus pallidus (medial)	178 ± 102* (10)	317 ± 48* (40)	10 ± 2* (10)	3.1 ± 1.9 (10)
Substantia nigra (pars compacta)	158 ± 41* (24)	229 ± 71* (15)		
Substantia nigra (pars reticulata)	86 ± 25* (25)	230 ± 54* (16)	25 ± 3* (10)	2.8 ± 1.6 (10)

For control values compare with Tables 1 and 2. *p < 0.05 compared to control values.

230

Table 1 and 2 for control values) show that there are no substantial changes in the peptide content of the frontal cerebral cortex (area 10), or the caudate nucleus and putamen. However, outside the caudate-putamen area the globus pallidus and substantia nigra are substantially depleted of CCK, methionine-enkephalin and substance P. (There were no significant changes in VIP content in any brain area). The reductions in CCK, substance P and methionine-enkephalin content are in areas in which the known pathology in Huntington's disease is less evident, whereas in the caudate-putamen in which there is between 50-80% loss of small (output?) neurones (Bruyn 1968, Lange et al. 1976) there are no significant changes in neuropeptide content. The cause of this apparent anomaly is probably the tissue shrinkage which occurs in the caudate nucleus and putamen which may compensate for the neuronal loss, whereas in areas to which these caudate-putamen neurones project, that is the substantia nigra and globus pallidus the loss of terminals and axons is not masked by tissue shrinkage and neuronal loss.

This data (Table 3) allows us to suggest that the human brain like the rat brain, contains neuronal cell bodies in the caudate-putamen complex which contain substance P, methionine-enkephalin and probably cholecystokinin immunoreactivity and project to the globus pallidus and substantia nigra. This loss of peptide neurones in Huntington's disease would be in addition to the loss of striatal γ-aminobutyric acid containing neurones in Huntington's disease (Bird and Iversen 1977) indicates that Huntington's disease probably involves a relatively general loss of output neurones from the striatum.

Conclusions

Our data indicate that significant amounts of several neuropeptides can be measured in post-mortem human brain. These peptides seem to be remarkably stable post-mortem and studies of brains from patients with Huntington's disease indicate that the observed neuropeptide changes reflect the known neuropathology.

Acknowledgements
   This work could not have been carried out without
the help of psychiatrists and neuropathologists
throughout the United Kingdom who assisted in the
collection of post-mortem brains, as well as Professor
Austin Gresham and members of staff in the Department
of Morbid Pathology at Addenbrooke's Hospital,
Cambridge. We acknowledge financial support from the
Danish Medical Research Council to Professor Rehfeld
and Dr Fahrenkrug and a twinning grant from the
European Brain and Behaviour Programme.

REFERENCES

Bird, E.D. and Iversen, L.L. (1974). Huntington's
      chorea - post-mortem measurement of glutamic acid
      decarboxylase, choline acetyltransferase and
      dopamine in basal ganglia. Brain 97, 457-472.
Bruyn, G.W. (1968). Huntington's chorea: historical,
      clinical and laboratory synopsis. In Handbook of
      Clinical Neurology, Vol. 6, Diseases of the
      Basal Ganglia, (eds. P.J. Virken and G.W. Bruyn),
      pp 298-378, North-Holland Publishing Co.,
      Amsterdam.
Cuello, A.C. (1978).   Endogenous opioid peptides in
      neurons of the human brain. Lancet ii, 291-293.
Emson, P.C. (1979).   Peptides as neurotransmitter
      candidates in the CNS. Progress in Neurobiology
      13, 61-116.
Emson, P.C., Arregui, A., Clement-Jones, V., Sandberg,
      B.E.B. and Rossor, M. (1980a). Regional distri-
      bution of methionine-enkephalin and substance P-
      like immunoreactivity in normal human brain and
      in Huntington's disease. Brain Res., (in press).
Emson, P.C., Fahrenkrug, J. and Spokes, E.G.S. (1979a).
      Vasoactive intestinal polypeptide (VIP): distri-
      bution in normal human brain and in Huntington's
      disease. Brain Res., 173, 174-178.
Emson, P.C. and Hunt, S.P. (1980). Anatomical
      chemistry of the cerebral cortex. In The Cerebral
      Cortex. (eds. F. Schmitt, F.G. Worden and
      S.G. Dennis), MIT Press (in press).
Emson, P.C., Hunt, S.P., Rehfeld, J.F. and Fahrenkrug,
      J. (1979b). Neurochemical studies on several
      brain peptides. In Molecular Endocrinology 1,
      (eds. J. MacIntyre and M. Szelke), Elsevier/
      North-Holland, Amsterdam.
Emson, P.C., Rehfeld, J.F., Langevin, H. and Rossor,
      M. (1980b). Reduction in cholecystokinin-like

immunoreactivity in the basal ganglia in Huntington's
disease. Brain Res., (in press).

Fahrenkrug, J. (1979). Vasoactive intestinal polypeptide: measurement, distribution and putative
neurotransmitter function. Digestion 19, 149-169.

Fahrenkrug, J. and Schaffalitzky de Muckadell, O.B.
(1977). Radioimmunoassay of vasoactive intestinal polypeptide (VIP) in plasma. J. Lab. Clin. Med
89, 1379-1388.

Gramsch, C., Höllt, V., Mehraein, P., Pasi, A. and
Herz, A. (1979). Regional distribution of meth-
ionine-enkephalin- and beta-endorphin-like
immunoreactivity in human brain and pituitary.
Brain Res., 171, 261-270.

Gramsch, C., Kleber, G., Höllt, V., Pasi, A., Mehraein,
P. and Herz, A. (1980). Pro-opicortin fragments
in human and rat brain: β-endorphin and α-MSH
are the predominant peptides. Brain Res., 192,
109-119.

Hong, J.S., Yang, H-Y.T. and Costa, E. (1977). On
the location of methionine enkephalin neurons in
rat striatum. Neuropharmacology 16, 451-453.

Innis, R.B., Correa, F.M.A., Uhl, G.R., Schneider, B.
and Snyder, S.H. (1979). Cholecystokinin
octapeptide-like immunoreactivity histochemical
localization in rat brain. Proc. natn. Acad.
Sci. USA 76, 521-525.

Kanazawa, I. and Jessell, T.M. (1976). Post-mortem
changes and regional distribution of substance P
in the rat and mouse nervous system. Brain Res.,
117, 362-367.

Kanazawa, I., Bird, E., O'Connell, R. and Powell, D.
(1977). Evidence for a decrease in substance P
content of substantia nigra in Huntington's
chorea. Brain Res., 120, 387-392.

Lange, H., Thorner, G., Hopf, A. and Schröder, K.F.
(1976). Morphometric studies of the neuropathological changes in choreactic disease. J. Neurol.
Sci., 28, 401-425.

Larsson, L-I., and Rehfeld, J.F. (1979). Localization
and molecular heterogeneity of cholecystokinin
in central and peripheral nervous system. Brain
Res., 165, 201-218.

Morris, H.R., Etienne, A.T., Dell, A. and Albuquerque,
R. (1980). High resolution purification of
neuropeptides. A rapid and specific method for
chemical characterisation. J. Neurochem., 34,
574-582.

Muller, J.E., Straus, E. and Yalow, R.S. (1977). Immu-
    nohistochemical localization in rabbit brain of a
    peptide resembling the COOH-terminal octapeptide
    of cholecystokinin. Proc. Nat. Acad. Sci. USA
    74, 3035-3037.
Pinget, M., Straus, E. and Yalow, R.S. (1978).
    Localization of cholecystokinin-like immunoreact-
    ivity in isolated nerve terminals. Proc. Nat.
    Acad. Sci. USA, 75, 6324-6326.
Rehfeld, J.F. (1978). Immunochemical studies on
    cholecystokinin II. Distribution and molecular
    heterogeneity in the central nervous system and
    small intestine of man and hog. J. Biol. Chem.,
    253, 4022-4030.
Snyder, S.H. (1980). Brain peptides as neurotransmitt-
    ers. Science (in press).
Spokes, E.G.S. (1980). Neurochemical alterations in
    Huntington's chorea: A study of post-mortem
    brain tissue. Brain 103, 179-210.
Spokes, E.G.S. and Koch, D.J. (1978). Post-mortem
    stability of dopamine, glutamate decarboxylase
    and choline acetyltransferase in the mouse brain
    under conditions simulating the handling of
    human autopsy material. J. Neurochem., 31,
    381-383.
Wesche, D., Höllt, V. and Herz, A. (1977). Radioimmuno-
    assay of enkephalin regional distribution in rat
    brain after morphine treatment and hypophysect-
    omy. Naunyn-Schmiedeberg's Arch. Pharmacol.,
    301, 79-82.

Section V
Pathological States
(B) Alzheimer's Disease and Senile Dementia

Biogenic Amines and Related Enzymes in Normal Aging, Senile Dementia and Chronic Alcoholism

BENGT WINBLAD, ROLF ADOLFSSON, STEN-MAGNUS AQUILONIUS,
ARVID CARLSSON, SVEN-ÅKE ECKERNÄS,
CARL-GERHARD GOTTFRIES, STEFAN MARKLUND,
AGNETA NORDBERG, LARS ORELAND, LARS SVENNERHOLM
and ÅSA WIBERG

Umeå Dementia Research Group, Department of Pathology,
University of Umeå, S-901 87 Umeå, Sweden

INTRODUCTION

When discussing aging and the aged it is of importance to separate normal or physiological aging or orthoinvolution from
pathological aging or pathoinvolution. The borderline between
normal and pathological aging, however, is not sharp. Normal
aging will usually not induce severe mental impairment if no
other complications arise, even if the individual reaches a very
old age.

Morphological investigations post-mortem have shown that the
degenerative changes observed even in normal aging do not affect
all parts of the brain to the same extent and are mostly located
to limbic structures and the cerebral cortex. Beside degenerative changes there is a neuronal and glial cell loss in the brain
cortex in normal aging (Brody, 1955; Henderson et al., 1980). It
was commonly assumed that the loss of cortical neurons is greater in senile dementia than in normal aging but quantitative
studies of cortex in senile dementia gave the somewhat surprising result that there is no significant loss of neurons nor
in glial cells in senile dementia (Terry, 1979). Structural
lesions in the dendritic tree with fewer and shorter terminal
segments, distortion and swelling of the cell-body dendrite
complex and progressive destruction of the dendrite domain might
contribute to the symptoms in the senile individual (Scheibel,
1979; Buell & Coleman, 1979).

There are many causes of the clinical syndrome dementia. Exclu-
ding patients with tumours, infection and dementia states
caused by vascular disease (multi-infarct dementia) there still
remains a large group of patients whose brains are atrophied
and display an excess of senile morphological changes such as

senile plaques, neocortical neurofibrillary tangles and hippo-campal granulovacuolar degeneration (Tomlinson et al., 1970; Corsellis, 1976). The diffuse brain atrophic process in senile dementia is considered to be irreversible and steadily progres-sive and hitherto no successful treatment seems to be available. The impairment of mental functions involves intellectual, emo-tional and motoric functions. There seems to be a relationship between cognitive decline on one side and number of senile plaques (Blessed et al., 1968), EEG-abnormalities (Obrist et al., 1963) and degree of atrophy on the other side (Willanger et al., 1968; Tomlinson et al., 1970; Huckman et al., 1975). In addition, good correlations have been found between cortical degenerative changes and regional cerebral blood flow (Gustafson et al., 1977) and also between EEG-abnormalities and regional cerebral blood flow (Johannesson et al., 1977).

Dementia disorders with onset before the age of 65 years are of-ten classified as presenile dementias or Alzheimer's disease, whereas when onset is after 65 years of age they are called senile dementia. Whether senile dementia is a homogenous or heterogenous group of dementias is still under discussion, and its distinction from presenile dementia may be impossible to prove. Although genetic investigations (Larsson et al., 1963)

	Normal aging	Senile dementia
Tyrosine hydroxylase	↓	↓↓
DOPA-decarboxylase	↓	↓↓
Dopamine betahydroxylase	↓	↓↓
Dopamine	↓	↓↓
Homovanillic acid	↓	↓↓
Noradrenaline	↓	↓↓
3-Methoxy-4-hydroxyphenylene-glycol (MHPG)	–	↑
5-Hydroxytryptamine (5-HT)	–	↓↓
5-Hydroxyindoleacetic acid (5-HIAA)	–	↓
Monoamine oxidase B (MAO-B)	↑	↑↑
Choline acetyltransferase (CAT)	↓	↓↓
Muscarinic binding sites	↓	–

Figure 1.   Summary of reported changes in transmitters and rela-ted enzymes in the brain in normal aging and dementia of Alzheimer type.

support the assumption that senile dementia and Alzheimer's disease are different types of disorders, structural as well as biochemical findings have often indicated that the two diseases should be brought together in one group called dementia of Alzheimer type.

Some post-mortem investigations of human brain tissue on the concentrations of the transmitter substances and related enzymes in normal aging and dementia of Alzheimer type are summarized in Figure 1.

The present investigation was performed in order to obtain more information about the levels of transmitters, their metabolites and related enzymes in normal aging, dementia of Alzheimer type and chronic alcoholism.

Methodological Aspects of Brain Post-Mortem Studies

In post-mortem investigations many factors have to be kept under careful control to obtain reliable results. These factors have earlier been discussed in detail (Adolfsson et al., 197´; Winblad et al., 1978). For example, both age and time between death and autopsy were found to have significant influence. Furthermore, the seasonal time of death as well as especially the clock time of death was found to be a significant factor for noradrenaline as well as for 5-hydroxytryptamine, 5-hydroxyindoleacetic acid, and, to a lesser extent dopamine concentrations (Carlsson et al., 1980).

Biochemical Findings in Normal Aging

Animal studies have shown that the syntheses of both noradrenaline and dopamine are reduced in the brains of senescent rats (Finch, 1973) as is the activity of tyrosine hydroxylase. The activity of tyrosine hydroxylase and DOPA decarboxylase have been examined post-mortem in human brain and significant age-related decreases in activity have been found (McGeer et al., 1971; McGeer and McGeer, 1973). Those reduced enzyme activities would suggest a reduced turnover of the catecholamines. Such a suggestion is in agreement with the finding of an age-related reduction in the concentration of dopamine in the human basal ganglia (Carlsson & Winblad, 1976). This finding was confirmed in another study by Adolfsson et al. (1979). A negative correlation between age and the concentrations of noradrenaline was also found in several brain regions, including basal ganglia (Winblad et al., 1978). The concentration of 5-hydroxytryptamine seems to be reduced with age in cortical areas but increased with age in the brain stem. Thus, the dopamine and noradrenaline transmitter systems seem selectively to be affected with age while the serotoninergic system is not affected, or at least not to the same extent. These changes may be coupled with altered sleeping habits and the increased frequency of depressive states in elderly.

Monoamine oxidase is a key enzyme in the degradation of monoamines (for review, see Tipton, 1973). The enzyme is thought to occur as two different forms in the brain, called monoamine oxidase -A and monoamine oxidase -B with different substrate and inhibitor specificities (for review, see Fowler et al., 1978). In a recent study a significant positive correlation of mono-amine oxidase -B with age in 19 out of 23 regions of the human brain was found, whereas no positive correlation with age was found for monoamine oxidase -A (Fowler et al., 1980). The increased monoamine oxidase -B activity was found to be due entirely to an increased enzyme concentration, rather than to an increased molecular turnover of the enzyme (Fowler et al., 1980). The activity of monoamine oxidase -B in the human platelets is not thought to increase with age (Murphy et al., 1976; Mann & Chiu, 1978) although there is a report to the contrary (Robinson et al., 1972).

In the cholinergic system the effect of age upon choline acetyl-choline acetyltransferase activity, acetylcholinesterase activity and the concentration of muscarinic receptor binding sites have been examined. An age-related reduction in the concentrations of muscarinic receptor binding sites has been observed by some investigators (White et al., 1977; Perry, 1980) but not by others (Davies & Verth, 1978). There appears to be a significant decrease of choline acetyltransferase activity with increasing age (McGeer & McGeer, 1975; Perry, 1980). Acetyl-cholinesterase does not appear to be affected by age in the range 60-90 years (Perry, 1980).

Biochemical Changes in Senile Dementia and Chronic Alcoholism
The established histopathological features of dementia of Alzheimer type have been known for a long time, whereas neuro-chemical research has mainly been undertaken in the last fifteen years. Although acetylcholine was the first neurotransmitter to be demonstrated (Loewi & Navratil, 1926) it has not been investigated in mental diseases to the same extent as other transmitters such as the catecholamines and serotonin. It was through the early studies by Pope et al. (1965) and Gott-fries et al. (1969; 1970; 1976) that it became clear that there were disturbances in the cholinergic as well as catecholaminer-gic and serotoninergic systems in dementia of Alzheimer type.

The results presented below are summarized from joint efforts by several Swedish laboratories. The brain regions were collected post-mortem in Umeå by Winblad. Autopsies were performed between 3 and 105 hours after death. Brain samples were taken from the hypothalamus, caudate nucleus, hippocampus and cortex gyrus cinguli immediately after removal from the skull. They were put into airtight packages and frozen at -70° C. The frozen samples were then pulverized in a mortar filled with liquid nitrogen and placed in plastic tubes and kept frozen. Controls were mainly vascular cases or malignancies without clinical or histopathological evidence of brain pathology. The diagnoses of the patients were undertaken by Adolfsson, Gottfries and Winblad. Patients with dementia of Alzheimer type were classified on clinical (insidious onset, memory loss, intellectual and personality deterioration, typical non-fluctuating progressive downhill course, absence of hypertension, stroke or other secondary causes of dementia; Roth 1978; Haase 1976; Isaacs 1979; Liston 1979) and on histopathological grounds (senile plaques, neurofibrillary tangles, granulovacuolar degeneration; Tomlinson et al., 1970).

Aliquots of the powders were distributed to several laboratories for analysis of monoamines and some of their precursors and metabolites (Carlsson), homovanillic acid and 3-methoxy-4-hydroxyphenylglycol (Svennerholm) monoamine oxidase -A and -B (Oreland, Wiberg), choline acetyltransferase (Aquilonius, Eckernäs), muscarinic receptors (Nordberg), and superoxide dismutase (Marklund).

TABLE 1. Age and time interval between death and autopsy[a]

	Controls (N=17)	Senile dementia (N=15)	Chronic alcoholics (N=14)
Age (years)	72.8 ± 1.90	75.7 ± 1.93	$52.8^{b} \pm 4.26$
Time between death and autopsy (hr)	31.0 ± 5.36	$53.2^{c} \pm 6.58$	$66.5^{b} \pm 8.05$

[a]Means ± SEM
[b]$p < 0.001$; [c]$p < 0.05$, two-tailed t-test

The mean values for age and post-mortem delay for the three groups are shown in Table 1. The age of controls and senile dementia cases was approximately the same, but the alcoholics were about 20 years younger. In most cases, this difference in age will either have no influence or increase the difference between controls and alcoholics.

The time between death and autopsy was, on the average, longer in the dementia and alcoholic groups than in the controls. In most cases, however, correction for this difference proved unnecessary and when done did not change the results in any important way. The uncorrected data are presented in Tables 2-5.

TABLE 2. The concentration of neurotransmitters and activities of neurotransmitter-related enzymes in hippocampus

Neurotransmitter or neurotransmitter-related enzyme under test	Controls (means ± SEM) (N=17)	Senile dementia (% control) (N=14)	Chronic alcoholics (% control) (N=13)
Homovanillic acid (nmoles/g)	2.0 ± 0.15	101	53^a
Norepinephrine (nmoles/g)	0.11 ± 0.009	58^a	70^b
MHPG (nmoles/g)	0.18 ± 0.021	128	111
5-HT (nmoles/g)	0.13 ± 0.016	22^c	47^a
5-HIAA (nmoles/g)	1.1 ± 0.08	64^a	59^c
MAO-B (d.p.m.)	3.266 ± 163	129^a	
CAT (pkatal/liter)	1.333 ± 115	57^a	81
Muscarinic receptor sites (pmol/mg protein)	128 ± 20.6	104	46^a

[a] $p < 0.01$; [b] $p < 0.05$; [c] $p < 0.001$ versus control, two-tailed t-test

TABLE 3. The concentration of neurotransmitters and activities of neurotransmitter-related enzymes in cortex gyrus cinguli

Neurotransmitter or neurotransmitter-related enzyme under test	Controls (means ± SEM) (N=17)	Senile dementia (% control) (N=15)	Chronic alcoholics (% control) (N=14)
Norepinephrine (nmoles/g)	0.15 ± 0.017	60^a	49^b
MHPG (nmoles/g)	0.22 ± 0.021	140^a	127
5-HT (nmoles/g)	0.06 ± 0.010	37^b	57
5-HIAA (nmoles/g)	0.61 ± 0.051	93	91
MAO-B (d.p.m.)	2.337 ± 121	117^a	
CAT (pkatal/liter)	776 ± 53	50^c	76^a

[a] < 0.05; [b] $p < 0.01$; [c] $p < 0.001$, two-tailed t-test

TABLE 4. The concentration of neurotransmitters and activities of neurotransmitter-related enzymes in hypothalamus

Neurotransmitter and neurotransmitter-related enzymes	Controls (Means ± SEM) (N=17)	Senile dementia (% control) (N=15)	Chronic alcoholics (% control) (N=14)
Dopamine (nmoles/g)	0.51 ± 0.079	56[a]	0[b]
Homovanillic acid (nmoles/g)	5.9 ± 0.52	83	85
Norepinephrine (nmoles/g)	6.6 ± 0.75	49	66
Normetanephrine (nmoles/g)	0.30 ± 0.044	58	80
MHPG (nmoles/g)	0.75 ± 0.090	87	103
5-HT (nmoles/g)	0.57 ± 0.078	46[c]	11[b]
5-HIAA (nmoles/g)	2.1 ± 0.21	69[d]	34[b]
MAO-B (d.p.m.)	4.734 ± 229	110	
CAT (pkatal/liter)	593 ± 77	74	113

[a] $p < 0.1$; [b] $p < 0.001$; [c] $p < 0.01$: [d] $p < 0.05$, two-tailed t-test

In dementia of Alzheimer type marked changes were found. For example, the level of dopamine in the caudate nucleus was 54 %, noradrenaline in the four regions 49-73 %, and 5-hydroxytryptamine 22 to 49 % of the respective control values. Choline acetyltransferase activity was 74 % of the controls in the hypothalamus but 50 to 57 % in the other three regions. The changes in the concentration of dopamine and 5-hydroxytryptamine were accompanied by reductions in the concentration of their respective metabolites, although these changes were less pronounced and did not reach significance in all cases. In two regions, 3-Methoxy-4-hydroxyphenyleneglycol levels were significantly increased in the senile demented. The multifactorial nature of dementia of Alzheimer type is supported by the observation that a complete separation between the dementia group and the controls was approached only when several of the examined transmitters and transmitter markers were taken into account simultaneously (Carlsson et al., 1979). The activity of monoamine oxidase -B was increased in the senile dementia group, while that of monoamine oxidase -A was unchanged (data not shown). Platelet monoamine oxidase activity was also found to be increased in a group of patients with dementia of Alzheimer type (table 6) (Adolsson et al., in press 1980a).

TABLE 5. The concentration of neurotransmitters and activities of neurotransmitter-related enzymes in caudate nucleus

Neurotransmitter and neurotransmitter-related enzymes	Controls (Means ± SEM) (N=17)	Senile dementia (% control) (N=15)	Chronic alcoholics (% control) (N=13)
Dopamine (nmoles/g)	15.8 ± 0.90^a	54^b	45^b
3-Methoxytyramine (nmoles/g)	6.6 ± 0.34^a	70^c	43^b
Homovanillic acid (nmoles/g)	22.7 ± 1.37	67^c	86
Norepinephrine (nmoles/g)	0.16 ± 0.012	73^d	65^d
MHPG (nmoles/g)	0.10 ± 0.019	183^d	164
5-HT (nmoles/g)	0.51 ± 0.043	49^b	30^b
5-HIAA (nmoles/g)	1.4 ± 0.11	87	60^c
MAO-B (d.p.m.)	3.390 ± 158	115^d	
CAT (pkatal/liter)	17.172 ± 1.879	55^c	90

^{a}N=54; $^b p < 0.001$ versus control; $^c p < 0.01$; $^d p < 0.05$, two-tailed t-test

TABLE 6. Activity of MAO (Mean ± SEM) in platelets estimated with two substrates in institutionalized Alzheimer patients and in controls matched for age and sex. The activity is expressed in nmoles per platelet per min $\cdot 10^9$.

	Alzheimer patients (N=11)	Controls (N=11)	Statistics[1]
MAO substrate			
β-phenethylamine	0.74 ± 0.07	0.53 ± 0.05	$p < 0.05$
Tryptamine	0.28 ± 0.04	0.20 ± 0.02	n.s.

[1]Wilcoxon's paired sample rank sign test

The finding of an increased platelet monoamine oxidase activity
may indicate that Alzheimer's disease is a generalized phenomenon not only confined to the brain (Winblad et al., 1978)
although results from other tissues are necessary to confirm
such a view. The relative increase in monoamine oxidase -B
activity in brain tissue both in normal aging and in dementia
of Alzheimer type may indicate that the underlying mechanisms
are similar.

In chronic alcoholics, similar changes in the concentrations of
the neurotransmitters were found, although in certain instances
(e.g. 5-hydroxytryptamine and 5-hydroxyindoleacetic acid in
hypothalamus and caudate nucleus) increases were more pronounced
and in others (e.g. choline acetyltransferase) less marked than
in the senile dementia cases (Tables 2-5). A significant decrease in choline acetyltransferase activity was found only in one
brain region of the alcoholics, i.e. the gyrus cinguli.

No significant difference in the concentration of muscarinic
binding sites was found in the hippocampus of dementia patients
with respect to controls, whereas in the chronic alcoholic group
a reduced concentration was found (Table 2).

As can be seen from Table 7 no decreased superoxide dismutase
activity was found, on the contrary, an increased activity was
found in the caudate nucleus and the hippocampus from the
Alzheimer brains.

TABLE 7. Superoxide dismutase activity (arbitrary units) in
          different brain parts.

	Cu Zn/mg protein		Mn/mg protein	
	Controls	Senile dementia	Controls	Senile dementia
Nucl. caudatus	545	671^X	39.7	43.0
Hippocampus	548	523	27.7	33.0^X
Cortex g. cinguli	503	563	33.0	36.2
Hypothalamus	584	684	42.1	44.2

^{X}significantly different from controls (t-test)

When the concentrations of amines or activities of enzymes were
plotted against age of the demented patient it appeared that the
patients with an early onset of the disease differed biochemically from those with a later onset. As can be seen in Figure 2,
for example, the activity of monoamine oxidase -B appears to be

relatively higher in the early onset dementia group than in the later onset group. The early onset group also seemed to have a lower choline acetyltransferase activity (Figure 3). These data are only preliminary and include only three demented patients above 75 years. It is possible, however, that the younger patients indeed constitute a more homogenous group and should be kept apart from those with a later onset of the disease.

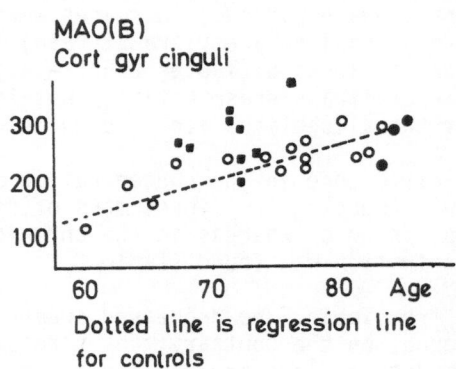

○ Controls
● Senile dementia
■ Alzheimer patients

MAO(B)
Cort gyr cinguli

Dotted line is regression line
for controls

Figure 2.

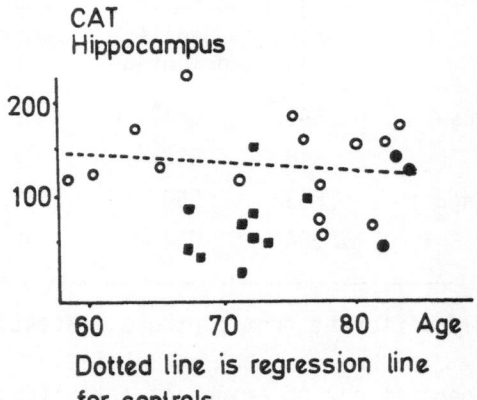

○Controls
●Senile dementia patients
■Alzheimer patients

CAT
Hippocampus

Dotted line is regression line
for controls

Figure 3.

In addition to our results described above, the activity of DOPA decarboxylase also seems to be affected in senile dementia. Bowen et al. (1974) have reported a reduced activity in cortical as well as subcortical brain structures in senile dementia. However, these findings were later questioned (Bowen et al., 1976). In the present CINP-congress there is a report of a reduced activity of dopamine β-hydroxylase in dementia of Alzheimer type (Cross et al.,1980).

A "cholinergic hypothesis" of dementia of Alzheimer type, in which functional abnormalities are postulated to be associated with defective cholinergic neurotransmission, has been suggested after the findings of significantly reduced brain activities of choline acetyltransferase (Davies & Maloney, 1976; Perry et al., 1977; White et al., 1977) and acetylcholinesterase (Davies & Maloney, 1976; Perry et al., 1978). The postsynaptic muscarinic receptor binding, on the contrary, has been reported to be unaffected (Perry et al., 1977; White et al., 1977; Davies & Verth, 1978; Nordberg et al., 1980). Pedigo et al. (1980), however, have reported a significant decrease in muscarinic receptor binding sites in the hippocampus in dementia of Alzheimer type similar to those found in brains from chronic alcoholics (Antuono et al., 1980; Nordberg et al., 1980).

The Newcastle group (Perry et al., 1978) have in their extensive post-mortem studies found that choline acetyltransferase and acetylcholinesterase activities in the brain decreased significantly as the mean plaque count rose. Furthermore, these authors found that in the demented the reduction in choline acetyltransferase activity correlated with degree of intellectual impairment. The unchanged muscarinic receptor binding and the reduced choline acetyltransferase activity could be explained by a degeneration of the presynaptic part of the neuron. Our recent studies (Bucht et al., 1979; Adolfsson et al., 1980 in press) show that patients with dementia of Alzheimer type have significantly lower fasting blood sugar (4.34 nmol/l) than in non-demented patient groups. Furthermore, the numerical values of the areas under the glucose tolerance test curves were also significantly lower in the dementia group when compared with the other groups. This result is in line with our earlier statement that dementia of Alzheimer type is a generalized disease not only confined to brain tissue (Winblad et al., 1978). Bowen et al. (1979) found a reduction in the activities of glycolytic enzymes in brain tissue of patients with Alzheimer's disease. It has been suggested (Perri et al., 1970) that, for the peripheral nervous system, a very slight decrease of the glucose levels leads to a rapid exhaustion of the neuronal firing. In addition, animal studies have shown that a slight hypoglycemia reduces the acetylcholine synthesis in the brain (Gibson & Blass, 1976). Whether or not the decreased blood glucose levels found in patients with dementia of Alzheimer type (Bucht et al., 1979; Adolfsson et al., 1980, in press) could impair the acetylcholine synthesis to the extent found by different independent research groups in brains from demented patients remains to be seen.

Conclusion: Evidence is accumulating that dementia of the Alzheimer type is a generalized disease not only confined to brain tissue. Due to the clinical symptoms and widespread neuronal degeneration in the brain it is unlikely that a single neurotransmitter would selectively be affected. The concentrations of dopamine, noradrenaline and 5-hydroxytryptamine were all reduced in senile dementia, with a concomitant increase in the activity of monoamine oxidase -B but not monoamine oxidase -A. The activity of choline acetyltransferase is also reduced. The data suggest that the process of neuronal aging may be accelerated in senile dementia. In chronic alcoholics, changes similar to those in senile dementia were observed, with respect to the concentrations of monoamines and the activity of choline acetyltransferase. For the future it seems important to identify the "biochemical profile" of the demented patient to get more positive results in substitution therapy trials than so far obtained.

Acknowledgement: This investigation was supported by grants from the Swedish Medical Research Council (12X-5664), Svenska Livförsäkrings Bolags Nämnd, Hansson's, Mångberg's, Osterman's and Pfannenstill's funds.

REFERENCES

Adolfsson, R., Gottfries, C.G. and Winblad, B. (1976). Methodological aspects of post-mortem investigations of human brain - with special reference to monoamines and related enzymes. In Proceedings of the Tenth Congress of the Collegium Intern. Neuro-Psychopharmacologicum, Quebec July 1976, (eds. P. Deniker, C. Radouco-Thomas and A. Villeneuve), Pergamon Press, Oxford, New York.

Adolfsson, R., Gottfries, C.G., Roos, B.E. & Winblad, B. (1979). Changes in the brain catecholamines in patients with dementia of Alzheimer type. Br. J. Psychiatry, 135, 216-223.

Adolfsson, R., Gottfries, C.G., Oreland, L., Wiberg, A. & Winblad, B. (1980a). Increased activity of brain and platelet monoamine oxidase in dementia of Alzheimer type. Life Sci. 27, 1029-1034.

Adolfsson, R., Bucht, G., Lithner, F. & Winblad, B. (1980b). Hypoglycemia in Alzheimer's disease. Acta Med.. Scand., in press.

Antuono, P., Sorbi, S., Bracco, L., Fusco, T. & Amaducci, L. (1980). Study of the cholinergic system in the ageing brain by means of discrete dissection technique. Results in senile

dementia of the Alzheimer type and alcoholic dementia.
In Ageing of the Brain and Dementia. Aging Vol. 13. (ed. L.
Amaducci, A. Davison and P. Antuono). Raven Press, New York.
pp 151-158.

Blessed, G., Tomlinson, B.E. & Roth, M. (1968). The association
between quantitative measures of dementia and of senile
changes in the cerebral grey matter of elderly subjects. Br.
J. Psychiatry, 114, 797-811.

Bowen, D.M., Flack, R.H.A., White, P., Smith, C.B. & Davison,
A.N. (1974). Brain-decarboxylase activities as indices of
pathological change in senile dementia. Lancet, 1, 1247-1249.

Bowen, D.M., Smith, C.B., White, P. & Davison, A.N. (1976).
Neurotransmitter-related enzymes and indices of hypoxia in
senile dementia and other abiotrophies. Brain, 99, 459-496.

Bowen, D.M., Spillane, J.A., Curzon, G., Meier-Ruge, W., White,
P., Goodhardt, M.J., Iwangoff, P. & Davison, A.N. (1979).
Accelerated ageing or selective neuronal loss as an important
cause of dementia? Lancet, 1, 11-14.

Brody, H. (1955). Organization of the cerebral cortex, part III
(A study of ageing in the human cerebral cortex). J. Comp.
Neurol., 102, 511-556.

Bucht, G., Adolfsson, R., Lithner, F. & Winblad, B. (1979). Sub-
normala fasteblodsocker och lågt svar på oral glukosbelast-
ning hos patienter med demens av Alzheimer typ. Acta Soc.
Med. Suecanae 88, 619.

Buell, S.J. & Coleman, P.D. (1979). Dendritic growth in the aged
human brain and failure of growth in senile dementia. Science
206, 854-856.

Carlsson, A., Adolfsson, R., Aquilonius, S.M., Gottfries, C.G.,
Oreland, L., Svennerholm, L. & Winblad, B. (1979). Biogenic
amines in human brain in normal aging, senile dementia and
chronic alcoholism. In Ergot Compounds and Brain Function:
Neuroendocrine and Neuropsychiatric Aspects. (ed. M. Goldstein et al.), Raven Press, New York. 295-304.

Carlsson, A., Svennerholm, L. & Winblad, B. (1980). Seasonal
and circadian monoamine variations in human brains examined
post-mortem. Acta Psychiatr. Scand., suppl. 75-85.

Carlsson, A. & Winblad, B. (1976). Influence of age and time
interval between death and autopsy on dopamine and 3-methoxy-
tyramine levels in human basal ganglia. J. Neural. Transm.
38, 271-276.

Corsellis, J.A.N. (1976). In Greenfield's Neuropathology, (eds.
W. Blackwood & J.A.N. Corsellis), Arnold, Edinburgh. 796-848.

Cross, A., Crow, T.J., Perry, E.k. & Kimberlin, R.H. (1980). Brain
dopamine-β-hydroxylase (DBH) activity in Alzheimer's disease
(AD). Abstracts of the 12th CINP Congress, Göteborg, Sweden
22-26 June, p 118. Supplement to Progress in Neuro-Psycho-
pharmacology, Pergamon Press.

Davies, P. & Maloney, A.J.F. (1976). Selective loss of central
cholinergic neurons in Alzheimer's disease. Lancet 2, 1403.

Davies, P. & Verth, A.H. (1978). Regional distribution of the muscarinic acetylcholine receptor in normal and Alzheimer's type dementia brains. Brain Res., 138, 385-392.

Finch, C.E. (1973). Catecholamine metabolism in the brains of ageing male mice. Brain Res., 52, 261-276.

Fowler, C.J., Callingham, B.A., Mantle, T.J. & Tipton, K.F. (1978). Substrate-selective activation of rat liver mitochondrial monoamine oxidase by oxygen. Biochem. Pharmac., 27, 97-101.

Fowler, C.J., Wiberg, A., Oreland, L., Marcusson, J. & Winblad, B. (1980). The effect of age on the activity and molecular properties of human brain monoamine oxidase. J. Neural Transm. in press.

Gibson, G.E. & Blass, J.P. (1976). Impaired synthesis of acetylcholine in brain accompanying mild hypoxia and hypoglycemia. J. Neurochem., 27, 37-42.

Gottfries, C.G., Gottfries, I. & Roos, B.E. (1969). The investigation of homovanillic acid in the human brain and its correlation to senile dementia. Brit. J. Psychiatry, 115, 563-574.

Gottfries, C.G., Gottfries, I. & Roos, B.E. (1970). Homovanillic acid and 5-hydroxyindoleacetic acid in cerebrospinal fluid related to rated mental and motor impairment in senile and presenile dementia. Acta Psychiatr. Scand., 46, 99-105.

Gottfries, C.G., Roos, B.E. & Winblad, B. (1976). Monoamine and monoamine metabolites in the human brain post-mortem in senile dementia. Aktuel. Gerontol. 6, 429-435.

Gustafson, L., Brun, A. & Ingvar, D.H. (1977). Presenile dementia: clinical symptoms, pathoanatomical findings and cerebral blood flow. In Cerebral Vascular Disease, (eds. J. Meyer, H. Lechner & M. Reivich), Excerpta Medica, Amsterdam. 5-9.

Haase, G.R. (1976). Diseases presenting as dementia. In Dementia, (ed. C.E. Wells & F.A. Davis), Philadelphia. 164-207.

Henderson, G., Tomlinson, B.E. & Gibson, P.H. (1980). Cell counts in human cerebral cortex in normal adults throughout life using an image analysing computer. J. Neurol. Sci., 46, 113-136.

Huckman, M.S., Fox, J. & Tropel, J. (1975). The validity of criteria for the evaluation of cerebral atrophy by computerized tomography. Radiology, 116, 85-92.

Isaacs. (1979). The evaluation of drugs in Alzheimer's disease. Age and Ageing, 8, 1-7.

Johannesson, G., Brun, A., Gustafson, L. & Ingvar, D.H. (1977). EEG in presenile dementia related to cerebral blood flow and autopsy findings. Acta Neurol. Scand., 56, 89-103.

Larsson, T., Sjögren, T. & Jacobson, G. (1963). Senile dementia: A clinical, sociomedical and genetic study. Acta Psychiatr. Scand., Suppl. 167.

Liston, E.H. (1979). Clinical findings in presenile dementia. A report of 50 cases. J. Nerv. Ment. Dis., 167, 337-342.

Loewi, O. & Navratil, E. (1926). Uber humorale übertragbarheit der Herznervenwirkung. X. Mitteilung. Uber das Schichsel des Vagnstoffs. Pflügers Arch. Ges. Physiol. 214, 678-688.

Mann, J. & Chiu, E. (1978). J. Neurology, Neurosurgery and Psychiatry, 41, 809-812.

McGeer, E.G. & McGeer, P.L. (1973). Some characteristics of brain tyrosine hydroxylase. In New Concepts in Neurotransmitter Regulation. (ed. A.J. Mandell), Plenum Press, New York. 53-69.

McGeer, E.G. & McGeer, P.L. (1975). Age changes in the human for some enzymes associated with metabolism of catecholamines, GABA and acetylcholine. In Neurobiology of Aging, (eds. J.N. Ordy & K.R. Brizzee), Plenum Press, New York. 287-305.

McGeer, E.G., McGeer, P.L. & Wada, S.A. (1971). Distribution of tyrosine hydroxylase in human and animal brain. J. Neurochem. 18, 1647-1658.

Murphy, D.L., Wright, C., Buchsbaum, M., Nichols, A., Costa, J.L. & Wyatt, R.J. (1976). Biochem. Med., 16, 254-265.

Nordberg, A., Adolfsson, R., Aquilonius, S.M., Marklund, S., Oreland, L. & Winblad, B. (1980). Brain enzymes and acetylcholine receptors in dementia of Alzheimer type (DAT) and chronic alcohol abuse. In Ageing of the Brain and Dementia. Aging Vol. 13. (ed. L. Amaducci, A. Davison and P. Antuono). Raven Press, New York. pp 169-172.

Obrist, W.D., Sokoloff, L., Lassen, N.A., Lane, M.H., Butler, R.N. & Feinberg, J. (1963). Relation of EEG to cerebral blood flow and metabolism in old age. EEG Clin Neurophysiol., 15, 510-519.

Pedigo, N.V., Reisine, T.D., Bird, E., Spokes, E., Enna, S.J. & Yamamura, H.Y. (1980). Neurotransmitter receptor alteration in Alzheimer's disease. In Ageing of the Brain and Dementia. Aging Vol. 13. (ed. L. Amaducci, A. Davison and P. Antuono). Raven Press, New York. p 147.

Perri, V., Sacchi, O. & Casella, L. (1970). Q. J. Exp. Physiol., 55, 25-35.

Perry, E.K. (1980). The cholinergic system in old age and Alzheimer's disease. Age and Ageing, 1, 1-8.

Perry, E.K., Gibson, P.H., Blessed, G., Perry, R.H. & Tomlinson, B.E. (1977). Neurotransmitter enzyme abnormalities in senile dementia. J. Neurol. Sci., 34, 247-265.

Perry, E.K., Tomlinson, B.E., Blessed, G., Bergmann, K., Gibson, P.H. & Perry, R.H. (1978). Correlation of cholinergic abnormalities with senile plaques and mental test scores in senile dementia. Brit. Med. J., 2, 1457-1459.

Pope, A., Hess, H.H. & Lewin, E. (1965). Trans. Amer. Neurol. Assoc. 89, 15.

Robinson, D.S., Nies, A., Davis, J.N., Bunney, W.E., Colburn, R.W., Bourne, H.R., Shaw, D.M. & Coppen, A.J. (1972). Lancet, 1, 290-291.

Roth, M. (1978). Diagnosis of senile and related forms of dementia. In Alzheimer's Disease: Senile Dementia and Related Disorders (Aging, Vol 7, ed. R. Katzman, R.D. Terry and K.L. Bick) Raven Press, New York. 7, 71-85.

Scheibel, A. (1979). The hippocampus: Organizational patterns in health and disease. Mechanisms of ageing and development, 9, 89-102.

Terry, R.D. (1979). Aging of the central nervous system and senile dementia. In Muscle, Nerve and Brain Degeneration. (ed. A.O. Kidman and J.K. Tomkins) Excerpta Medica. Amsterdam-Oxford. pp 187-202.

Tipton, K.F. (1973). Brit. Med. Bull., 29, 116-119.

Tomlinson, B.E., Blessed, G. & Roth, M. (1970). Observations on the brains of demented old people. J. Neurol. Sci., 11, 205-242.

White, P., Goodhardt, M.J., Keet, J.P., Hiley, C.R., Caraso, L.H., Williams, J.E.J. & Bowen, D.M. (1977). Neocortical neurons in elderly people. Lancet 1, 668-671.

Willanger, R., Thygesen, P., Nielsen, R. & Petersen, O. (1968). Intellectual impairment and cerebral atrophy. A psychological and neuroradiological investigation. Dan. Med. Bull., 15, 69-93.

Winblad, B., Adolfsson, R., Gottfries, C.G., Oreland, L. & Roos, B.E. (1978). Brain monoamines, monoamine metabolites and enzymes in physiological ageing and senile dementia. In Recent developments in mass spectrometry in biochemistry and medicine, (ed. A. Frigerio), Plenum Publ. Corp., New York. 1, 253-267.

Biochemical Changes in Alzheimer's Disease in Relation to Histopathology, Clinical Findings and Pathogenesis

D. M. BOWEN, N. R. SIMS and A. N. DAVISON

Department of Neurochemistry,
Institute of Neurology,
Queen Square,
London, WC1N 3BG

In Alzheimer's-type dementia (ATD) - used here to include both presenile and senile forms of the disease (as these appear to be indistinguishable on neuropathological grounds) - there is a global impairment of higher cortical function. Short-term memory is particularly affected. The natural history is frequently progressive and is irreversible. There is still some debate as to whether this condition represents a separate disease entity distinct from an exaggeration of normal brain senescence. In a number of other dementing disorders, degenerative changes have been attributed to a definitive cause. Examples are the transmissable agent of Creutzfeldt-Jacob disease and aluminium dialysis dementia. Alcoholism is also associated with brain degeneration and cognitive loss. Recent studies of ATD have attempted to define more closely the major changes associated with the disease, using biochemical and histological techniques.

NERVE CELL LOSS

Although decrease in brain weight is common in ATD, increased ventricular volume does not correlate well with the degree of dementia. Atrophy particularly affects the frontal and temporal lobes. A study in our laboratory indicated about a 20 per cent decrease in wet weight and protein content of the temporal lobe (Bowen et al. 1979) and a similar reduction has been reported for hemispheric volume (Miller et al. 1980). There has been debate as to whether such loss results from a reduction in neuronal population, or is due to shrinkage of nerve cell bodies and loss of dendrites. Quantitative histochemical data of de Kosky and Bass (1980) show that in the frontal association cortex of both aged and demented subjects, there was a 20 per cent loss in cell-packing density. Using histological methods, Colon (1973) found a 57 per cent loss of neocortical neurons in Alzheimer's disease; but others have found no change in cell population in the neo-cortex of ATD patients. This discrep ncy may reflect regional differences or may be due to a selective loss of large neurons whilst the number of smaller neurons remains unchanged (see Terry, 1979). Ball (1977) has demonstrated neuronal loss in the hippocampus which correlates with both the number of nerve cells with neuro-fibrillary degeneration and the number with granulovacuolar degeneration.

Since quantitative histological methods have some diffi-culties of interpretation, we decided to utilize biochemical markers of cell types and metabolism. For this purpose one intact temporal lobe was completely homogenized or finely minced and aliquots used for analysis. The alternate lobe was examined histologically. Some of the markers assayed had a good degree of specificity (e.g. DNA for cells or 2', 3'-cyclic nucleotide 3'-phosphohydrolase for myelin); others had less precise localization but were more concentrated in one cell type or component than another (e.g. carbonic anhydrase in glial cells or ATPase are in cell membranes). In the first studies by Bowen and his colleagues (Bowen et al. 1977), biochemical measurements were made on whole temporal lobe homogenates from cases of his-tologically verified ATD, without and with cerebro-vascular disease (mixed senile and vascular dementia). The data suggested significant loss of nerve cell components. In a later investiga-tion (Bowen et al. 1979), results on 35 biochemical constituents were assessed in 17 histologically verified ATD cases, and the findings were compared with those from 11 with mixed senile and vascular dementia and with 16 normal age-matched controls (81 $\pm$ 15 years old). In non-vascular ATD, 19 biochemical markers were significantly reduced compared to only 10 affected in cases with mixed pathology. The reductions in ganglioside content and activity of a myelin marker in ATD are consistent with about a 30 per cent loss or shrinkage of nerve cells. Since relatively younger patients (70 to 80 years old) compared to elderly are most affected, the data also suggested that ATD is a distinct disease process superimposed on natural aging. This is con-sistent with the conclusion of Colon (1973) from studies of cell loss.

Although neuropathological features of ATD can first be detected from 40 to 50 years onwards in intellectually preserved individuals, the proportion of cases where plaques, tangles or granulovacuoles can be detected remains low up to the eighth decade (Tomlinson, 1977). In some regions of the aging brain, examination by the Golgi method shows that there is loss of

254

dendritic processes and synaptic contacts (Schiebel & Tomiyasu, 1978). Although in their study Buell & Coleman (1979) saw gross-ly atrophic dendritic arborisation, the intensity of such change was no different in adult (mean age 51.2 years), compared to elderly (mean age 79.6 years) subjects. In the normal aging human parahippocampal gyrus, Buell & Coleman found continued dendritic arborisation. Indeed, the net length of individual terminal segments of the average apical tree increased by 0.21 µm/year from 44 to 92 years of age. This suggests unexpected plasticity within the aging brain; possibly this reactive synaptogenesis (Cotman & Scheff, 1979) is in response to neuro-nal loss. Buell & Coleman (1979) found that in patients with ATD dendritic trees were less extensive than in adult pyramidal neurons of the parahippocampal gyrus, as the terminal nerve cell segments were fewer and shorter. In the cingulate gyrus and the hippocampus, a decrease in the number of dendrites and in the extent of dendritic arborization was observed in ATD by Mehraein et al. (1975). The loss was greatest in the younger group of patients.

CHARACTERISTIC HISTOPATHOLOGICAL CHANGES
In ATD widespread characteristic neuropathological changes are seen - neuritic or senile plaques with an amyloid core, tangles of paired helical filaments (originating from uniden-tified endogenous brain proteins), and in the hippocampus granu-lovacuolar degeneration of neurons.

Light microscopical and ultra-structural studies demon-strate that the senile plaques consist of an amyloid core with degenerating neurites and presynaptic terminals. The plaques are well defined, small patches of altered tissue occurring mainly in the cortex, but also in the amygdaloid nucleus and infrequently in the striatum, thalamus and brain stem. The centre or 'core' of the plaque is possibly derived from IgG (Ishii & Haga, 1976), suggesting that immunological factors are involved in pathogenesis. A localized increase in the activity of oxidative enzymes in senile plaques (Friede, 1965) is probably due to the projection of processes from astrocytes. There is an unexplained finding of a high silicon concentration in the 'cores' and 'rims' of plaques (Nikaido et al., 1971).

Granulovacuoles, which are confined to the pyramidal cells of the hippocampus, are like senile plaques and tangles in that they increase in number with advancing years in non-demented subjects (Ball & Vis, 1979). Granulovacuoles show a positive reaction to acid phosphatase and other hydrolases (Krigman et al. 1965), which is indicative of a change in neuronal lysosomes.

Hippocampal neurones are most likely to show tangles or neurofibrillary degeneration, but they occur also in the neo-

255

cortex and sometimes in the brain stem. Electron microscopic studies on biopsy samples of neocortex led Kidd (1963, 1964) to describe the change as being paired helical filaments, and this has been confirmed (Wisniewski, Narang & Terry, 1976). Although the paired helical filaments have been described only in human nerve cells, they occur in the brain of mentally normal elderly subjects and in conditions such as mongolism, post-encephalitic Parkinsonism, brain damaged boxers and the amytrophic lateral sclerosis Parkinsonism dementia complex. Recent studies suggest that the protein component of tangles is antigenically related to a fibrous protein that normally occurs in brain, but the identity of the precursor is unclear (Grundke-Iqbal et al. 1979; Gambetti et al. 1980).

PATHOGENESIS

A number of factors have been suggested as having a role in the development of ATD, but currently there is no strong evidence favouring any of these. Mirjakawa et al. (1974) reported that senile plaques are often associated with capillaries, perhaps indicating that a blood-borne factor such as a virus or toxic agent may be involved in the disease. Although a dementia associated with a transmissable slow virus is found in Creutz-feldt-Jacob disease, this is a rare condition and there is little evidence for the involvement of such an agent in ATD (Corsellis, 1979; Whalley et al. 1980). De Boni and Crapper (1978) reported a heat labile high molecular weight factor in extracts of brain from ATD patients which induces paired helical filament formation in human embryonic brain grown in tissue culture. The paired helical filament assembly factor has properties distinct from either a conventional or unconventional virus of the scrapie type. This could lead to an impairment in slow cytoplasmic transport, and result in a reduction in synaptic activity.

Aluminium has been suggested as a possible neurotoxic agent as increased concentrations have been found in the brain of individuals with dementia (Crapper et al. 1976). The majority of brain aluminium is localised in the nucleus and studies of twelve cases of ATD showed an increase of intranuclear concentrations. This was most marked in presenile cases, these having about twice the amount of age-matched controls. Curiously, less nuclear aluminium is found in the brain of cases of dialysis dementia, even though the whole tissue aluminium content was between 10 to 27 μg/g dry weight, compared to about 1.8 μg/g in the normal human cortex. Some evidence was found by Crapper et al. (1980) for an association between increased aluminium and neurofibrillary tangle formation. It is, therefore, of interest that Perl & Brody (1980) found, using the scanning electron microscope with x-ray spectrometry, that only neurons with neuro-fibrillary degeneration contain detectable amounts of nuclear aluminium. This was true for ATD and non-demented senile cases. This may simply reflect accumulation of aluminium in a failing senescent nerve cell, rather than indicate that the metal is the primary pathological agent.

Oxygen deprivation may be another factor in ATD for the pattern of hippocampal damage is similar in hypoxic and Alzheimer brain (Ball, 1977, 1978). Plum (1978), however, found little evidence to support the view that chronic hypoxia underlies the brain changes of Alzheimer's disease.

It has been suggested that a genetic defect may predispose individuals to ATD, although direct evidence of familial involvment is lacking in the majority of cases. In normal aging there is a loss of RNA and a smaller nucleolus and this is reported to also occur in ATD (Mann & Yates, 1979). But in ATD the loss of RNA is said to be more widespread. This is most marked in the hippocampus but a 30 per cent loss of RNA is found in such diverse neurons as Purkinje cells, the dentate, the inferior olives and cortical Betz cells (Yates et al. 1980). Crapper-McLachlan & De Boni (1980) have found that the proportion of euchromatin is about 75 per cent of nuclear DNA in multi-infarct dementia and elderly controls, whereas it is reduced to an average of 55 per cent in ATD. The reduction is similar in both isolated glia and neurons. It seems, therefore, that a possible defect in transcription may be associated with a non-specific decrease in protein synthesis in ATD, for no abnormal protein has been detected. However, direct measurement in a microsomal preparation from neocortex (Suzuki et al. 1964) did not show changes in protein synthesis from control values. Further studies of neuron-specific, protein synthesis are needed.

BRAIN METABOLISM

In patients described as having "chronic brain syndrome and psychosis" the early studies of Kety demonstrated significant reduction in cerebral oxygen and glucose consumption (Sokoloff, 1976). Data on regional cerebral blood flow (rCBF) and in vivo rates of glucose and oxygen utilization are available for only groups of presumptive cases of ATD (i.e. for most cases the clinical diagnosis has not been confirmed by a neurohistological examination). Thus Raiche et al. (1978) have reported a one-third reduction in rCBF in demented patients with cerebral atrophy and a comparable loss of oxygen utilization by the brain. Most studies agree that there is a reduction in rCBF in at least the later stages of presenile and senile dementia. However, the reductions of rCBF over the slight decrease due to normal aging (Lavy et al. 1978) have been found by Melamed et al.(1978) not to correlate with either ventricular enlargement or widening of cortical sulcii. It was therefore suggested that loss of brain substance is not an important factor in reducing rCBF. Evidence that changes in rCBF may depend upon functional demand has been provided by Ingvar, Larsen and associates (Ingvar et

257

al. 1978). They find that the augmentation of flow in demented
patients in the associated cortex during attempted activation
by reading and psychological tests was less than normal, and
in some cases rCBF was even diminished.

Glucose utilization has been measured in vitro, with
samples of neocortex removed at diagnostic craniotomy from ATD
cases. Incubations were performed in both low (5 mM) and high
(31 mM) potassium concentrations, thus providing measurements
under resting and stimulated conditions. $U-^{14}C$ glucose was
used as the radioactive precursor and its utilization was measured by trapping the $^{14}CO_2$ evolved. The $^{14}CO_2$ production was
not significantly affected (Sims et al. 1980). This result is
consistent with previous data which showed that measurements of
oxygen uptake and lactic acid production in slices of a cerebral
cortex biopsy sample from a patient with ATD were similar to
those of a control sample (Suzuki et al. 1965).

DEFECT IN SYNAPTIC TRANSMISSION

Since histopathological observations indicate that abnormal
and probably reduced numbers of synaptic terminals are a feature
of ATD, interest has focused on chemically-mediated synaptic
transmission. It is generally not practical to determine the
concentration of particular neurotransmitters in post-mortem
brain, but the activity of the appropriate enzyme serves to
indicate synthetic potential. Preliminary examination (Bowen
et al. 1974) of post-mortem tissue from demented patients suggested that a possible reduction in glutamate decarboxylase and
aromatic amino acid decarboxylase (DOPA decarboxylase) activity;
but later work showed that both enzymes were affected by the
patient's terminal state. This conclusion was confirmed by
our finding that glutamate decarboxylase activity is normal in
biopsy tissue from ATD cases (Spillane et al. 1977). Reductions
have been seen in the concentration of the dopamine metabolite
homovanillic acid (HVA) in the CSF (Gottfries et al. 1969), and
in HVA, dopamine and noradrenaline concentration in some areas
of the brain of demented patients (Winblad et al. 1978). Changes
in HVA concentration in the CSF have not been observed by all
workers (Parkes et al. 1974). Yates et al. (1980) found no
change in the dopamine concentration in the caudate nucleus and
substantia nigra in ATD, compared with age-matched controls.
The dopaminergic system is unlikely to be directly involved in
ATD (see also Davies & Maloney,1976).

Changes in Presynaptic Cholinergic Activity in ATD

Using post-mortem tissue from patients with ATD and suitable
age-matched controls, Bowen et al. (1976) found that there was
a reduction in choline acetyltransferase (CAT) activity, which
was inversely correlated with the degree of neuropathological
damage in the pre-frontal cortex. CAT is more resistant to
hypoxic change than other transmitter synthesizing enzymes.
Davies & Maloney (1976), and Perry et al. (1977a) in cases of

ATD, showed that reduction in CAT activity was even more marked in the hippocampus and mamillary body. In order to check that such decreases in enzyme activity were not due to the terminal condition of the patient or to post-mortem artefact, we have measured CAT activity in brain biopsy samples (Bowen et al.1979). In comparison with controls, biopsy samples with histological evidence of ATD were also found to have a decrease in the mean CAT activity. Perry et al. (1978) have extended the earlier observations (Bowen et al. 1976) and found that there is a decrease in CAT and acetylcholinesterase activity in the cerebral cortex, which inversely relates to the degree of neuropathological change in the brain, as measured by senile plaque count. The reduction in CAT activity correlated with the extent of intellectual impairment measured, using a memory impairment test within six months prior to death (Perry et al. 1978). Although these observations suggest a reduction in presynaptic cholinergic activity, no changes have been found in the concentration of post-synaptic muscarinic cholinergic binding sites in cortical tissue, or in the total receptor protein in the whole temporal lobes (White et al. 1977; Bowen et al. 1979; Perry et al. 1978). However, the high affinity binding of LSD, a potential marker of serotoninergic receptors, and a marker of serotonin turnover (ratios of the concentrations of 5-hydroxyindole acetic acid and serotonin), are reduced in dement temporal lobe (Bowen et al. 1979; see also Carlsson et al. 1980). Further work is needed, for the latter change may be related to terminal phenomena, but LSD binding was unaltered in the caudate nucleus, which is of interest, as senile degeneration occurs infrequently in this region. The reduction in the markers of serotoninergic pathways were 60 to 70 per cent, while CAT was more reduced (to 35 per cent of control). In their study, **Reisine** et al. (1978) found that GABA binding was reduced in both cortex and caudate nucleus. Spiroperidol binding was unaffected in cortex. In the hippocampus, they found a significant reduction (60 per cent of control) in quinuclidinyl benzilate binding, while others (Davies & Verth, 1978) find no change.

In vivo and in vitro studies indicate that acetylcholine is derived from acetyl-CoA and choline through the action of CAT activity, an enzyme present in excess in the pre-synaptic terminal. Although low affinity transport of choline may contribute to uptake, a sodium dependent high affinity choline uptake system (Kuhar & Murrin, 1978) is probably primarily involved. This system is uniquely localized to cholinergic neurons and their nerve terminals. It appears that CAT activity is unlikely to be rate-limiting in acetylcholine formation (Marchbanks & Wonnacott, 1979). Sims et al. (1980) have adapted the methods of Gibson et al. (1975), to give a sensi-

tive method for measurement of acetylcholine synthesis in
brain tissue prisms prepared from biopsy samples of human neocortex. Fresh prisms (a preparation largely containing intact
nerve terminals) were incubated in a physiological medium with
U-^{14}C-glucose as substrate and glucose utilization (from radioactive CO_2) and ^{14}C acetylcholine synthesis were determined.
Biopsy samples from ATD patients were compared with tissue
removed in the course of neurosurgery from patients without
clinical features of dementia. The samples from ATD patients
showed a marked reduction in the acetylcholine synthesized
which correlated with the decrease in CAT. Assuming that the
enzyme is not rate limiting in acetylcholine synthesis in
human material, the parallel decrease in these two cholinergic
markers may reflect a loss of cholinergic neurons or terminals.
The cholinergic defect appeared to be related to both the psychological deficit and extent of atrophy (assessed by computerized
axial tomatography) in individual patients, suggesting that the
biochemical findings are of clinical significance.

Cholinergic Markers in Non-Dement Aging Brain
     The data for the control brains of Davies (1979), McGeer
and McGeer (1976) and Perry et al. (1977b) indicate that by
95 years CAT activity is only about 10-20% of the value for
neocortex from young adult brain. By contrast, neither Bowen
et al. (1976, 1979), White et al. (1977), Spokes (1979)
nor Carlsson et al.(1980) find significant change. Differences
in the various control groups may explain this discrepancy.
For example, those studies by Davies (1979), McGeer and McGeer
(197∴) and Spokes (1979) are described as non-neurological controls, whereas the cases of Bowen and Perry were selected primarily on the basis of mental and histological normality
and mental normality, respectively. Due to the difficulty of
obtaining specimens from proven mentally normal elderly subjects,
the analysis of animal brain may provide supportive data about
possible age-related changes in human brain. Thus, in preliminary studies, using rat and rabbit, neither Meek et al. (197⁻),
Lippa et al. (1979) nor Makman et al. (1979) found a significant
age-related change in cortical CAT activity. This is in accord
with the earlier data of McGeer et al. (1971) showing no significant change across age in rat brain minus caudate nucleus and
cerebellum.

     Two studies in Man provide evidence that muscarinic cho-
linergic receptors decline in normal old age. Firstly, blockade of central muscarinic receptors converts the pattern of
intellectual performance of young adults to that of elderly
subjects (Drachman & Leavitt, 1974). Secondly, a significant
age-related decline of 32% in muscarinic receptor binding

occurs in neocortex with little or no evidence of senile
degeneration from subjects aged 65-92 years (White et al. 1977).
Animal studies are also indicative of an age-related decline
in cholinoreceptive cells (Lippa et al. 1979; Strong et al.
1979; Freund, 1980; Morin & Westerlain, 1980). Loss of cholinoreceptive cells appears also to occur in the temporal lobe in
Pick's disease (White et al. 1977), a rare dementing condition
which is often impossible to distinguish from ATD using clinical
criteria. The binding of three muscarinic ligands has been
studied (atropine, scopolamine and quinuclidinyl benzilate) in
ATD, and all show lack of change in the temporal lobe and neocortex (White et al. 1977; Bowen et al. 1979; Perry et al.1977b;
Davies and Verth, 1978). Thus, in this respect the brain in ATD
differs from that in normal aging and Pick's disease.

CONCLUSION

Neurophysiological studies, such as rCBF measurements,
suggest that in ATD neuronal metabolism is altered. Despite
some reduction in glycolytic enzyme activity found in the temporal lobe of post-mortem ATD cases (Bowen et al. 1979), our
in vitro metabolic study shows that overall glucose utilization
is not significantly impaired (Sims et al. 1980). One interpretation of the apparent anomaly is that metabolic normality is
restored during the preparation of the tissue prisms.
The cholinergic system is apparently affected at an early stage
in the pathogenesis of ATD, for CAT activity is reduced in both
biopsy specimens and post-mortem material from less severely
impaired ATD patients. These observations, together with the
relationship between CAT activity, senile histological degeneration, and mental state, suggest that the depletion in activity is of clinical significance. The enzyme may not be critical in regulating acetylcholine synthesis, since studies on
animal brains indicate that it is present in excess. Thus, the
data (Sims et al. 1980) showing that acetylcholine synthesis is
reduced in ATD is an important finding. Since the incubations
were carried out in the presence of excess choline (2mM) this
indicates that choline therapy would be unlikely to reverse
the cholinergic deficit. Further work on autoreceptors
(Lefresne et al. 1978) and possible non-cholinergic sites on
cholinergic terminals, may lay the basis for rational drug
trials. Whether or not these provide effective therapy will
depend largely upon the extent to which other types of neuron
are affected.

REFERENCES

Ball, M.J. (1977). Neuronal loss, neuro-fibrillary tangles and granulovacuolar degeneration in the hippocampus with ageing and dementia. Acta Neuropath. (Berl.),37, 111-118.

Ball, M.J. (1978). Topographic distribution of neuro-fibrillary tangles and granulovacuolar degeneration in hippo-campal cortex of aging and demented patients. A quantitative study. Acta Neuropath. (Berl.), 42, 73-80.

Ball, M.J. and Vis, C.L. (1979). Relationship of granulo-vacuolar degeneration in hippocampal neurones to aging and to dementia in normal-pressure hydrocephalics. J. Gerontol., 33, 815-824.

Bowen, D.M., Flack, R.H.A., Smith, C.B., White, P. and Davison, A.N. (1974). Brain-decarboxylase activities as indices of pathological change in senile dementia. Lancet i, 1247-1249.

Bowen, D.M., Smith, C.B., White, P. and Davison, A.N. (1976). Neurotransmitter-related enzymes and indices of hypoxia in senile dementia and other abiotrophies. Brain, 99, 459-496.

Bowen, D.M., Smith, C.B., White, P., Flack, R.H.A., Carrasco, L., Gedye, J.W. and Davison, A.N. (1977). Chemical pathology of the organic dementias. II Quantitative estimation of cellular changes in post-mortem brains. Brain, 100, 427-453.

Bowen, D.M., White, P., Spillane, J.A., Goodhardt, M.J., Curzon, G., Iwangoff, P., Meier-Ruge, W. and Davison, A.N.(1979). Accelerated aging or selective neuronal loss as an important cause of dementia? Lancet i, 11-14.

Buell, S.J. and Coleman, P.D. (1979). Dendritic growth in the aged human brain and failure of growth in senile dementia. Science, 206, 854-855.

Carlsson, A., Adolfsson, R., Aquilonius, S.M., Gottfries, C.G., Oreland, L., Svennerholm, L. and Winblad, B. (1980). Biogenic amines in human brain in normal aging, senile dementia and chronic alcoholism. In Ergot Compounds and Brain Function: Neuroendocrine and Neuropsychiatric Aspects (eds. M. Goldstein, D.B. Calne, A. Lieberman and M.O. Thorner), Raven Press, New York, p. 295-304.

Colon, E.J. (1973). The cerebral cortex in pre-senile dementia. Acta Neuropath. (Berl.), 23, 281-290.

Corsellis, J.A.N. (1979). On the transmission of dementia. A personal view of the slow virus problem. Brit. J. Psychiat., 134, 553-559.

Cotman, C.W. and Scheff, S.W. (1979). Compensatory synapse growth in aged animals after neuronal death. Mechanisms of Ageing and Development, 9, 103-117.

Crapper, D.R., Krishnan, S.S. and Quittkat, S. (1976). Aluminium, neurofibrillary degeneration and Alzheimer's disease. Brain, 99, 67-80.

Crapper, D.R., Quittkat, S., Krishman, S.S., Dalton, S.J. and DeBoni, U. (1980). Intranuclear aluminium content in Alzheimer's disease, dialysis encephalopathy and experimental aluminium encephalopathy. Acta Neuropath. (Berl.), 50, 19-24.

Crapper-McLachlan, D.R. and DeBoni, U. (1980). Etiologic factors in senile dementia of the Alzheimer type. In Aging of the Brain (eds. L. Amaducci, P. Antuoni and A.N. Davison), Raven Press, New York, in press.

Davies, P. (1979). Neurotransmitter-related enzymes in senile dementia of the Alzheimer type. Brain Research, 171, 319-327.

Davies, P. and Maloney, A.J.F. (1976). Selective loss of central cholinergic neurons in Alzheimer's disease. Lancet ii, 1403.

Davies, P. and Verth, A.H. (1978). Regional distribution of muscarinic acetylcholine receptors in normal and Alzheimer-type dementia brain. Brain Research, 138, 385-392.

De Boni, V. and Crapper, D.R. (1978). Paired helical filaments of the Alzheimer type in cultured neurones. Nature, 271, 566-568.

Dekosky, S.T. and Bass, N.H. (1980). Effects of aging and senile dementia on the microchemical pathology of human cerebral cortex. In Aging of the Brain and Dementia (eds. L. Amaducci, P. Antuoni and A.N. Davison), Raven Press, New York, in press.

Drachman, D.A. and Leavitt, J. (1974). Human memory and the cholinergic system. Arch. Neurol. (Chicago), 30, 113-121.

Freund, G. (1980). Cholinergic receptor loss in brains of aging mice. Life Sci., 26, 371-375.

Friede, R.C. (1965). Enzyme histochemical studies of senile plaques. J. Neuropath. Exp. Neurol., 24, 477-491.

Gambetti, P., Velasco, M.E., Dahl, D., Bignami, A. and Roessmann, V. (1980). Neurofibrillary tangles in Alzheimer's disease. An Immunohistochemical study. In Aging of the Brain and Dementia (eds. L. Amaducci, P. Antuoni and A.N. Davison), Raven Press, New York, in press.

Gibson, G.E., Jope, R. and Blass, J.P. (1975). Decreased synthesis of acetylcholine accompanying impaired oxidation of pyrivic acid in rat brain minces. Biochem. J., 148, 17-23.

Gottfries, C.G., Gottfries, J. and Roose, B.E. (1969). The investigations of homovanillic acid in the human brain and its correlation to senile dementia. Brit. J. Psychiat., 115, 563-574.

Grundke-Iqbal, I., Johnson, A.B., Terry, R.D., Wisniewski, H.M. and Iqbal, K. (1979). Alzheimer Neurofibrillary Tangles: Antiserum and Immunological staining. Ann. Neurol., 6, 532-537.

Ingvar, D.W., Brun, A., Hagberg, B. and Gustafson, L.(1978). Regional cerebral blood flow in the dominant hemisphere in confirmed cases of Alzheimer's disease. In Alzheimer's disease: Senile dementia and related disorders (eds. R. Katzman, R.D. Terry and K.L. Bick), Raven Press, New York, p.441-451.

Ishii, T. and Haga (1976). Immunoelectron microscopic localisation of immunoglobulins in amyloid fibrils of senile plaques. Acta Neuropath. (Berl.), 36, 243-249.

Kidd, M. (1963). Paired helical filaments in electron microscopy of Alzheimer's disease. Nature, Lond., 197, 192-193.

Kidd, M. (1964). Alzheimer's disease - an electron microscopical study. Brain, 87, 307-315.

Krigman, M.R., Feldman, R.G. and Bensch, K. (1965). Alzheimer's presenile dementia. A histochemical and electron microscopic study. Lab. Invest. 14, 381-396.

Kuhar, M.J. and Murrin, L.C. (1978). Sodium-dependent high affinity choline uptake. J. Neurochem. 30, 15-21.

Lavy, S., Melamed, E., Bentin, S., Cooper, G. and Rinof, Y. (1978). Bihemisphere decreases of regional cerebral blood flow in dementia. Ann. Neurol., 4, 445-450.

Lefresne, P., Rospars, J.P., Beaujouan, J.C., Westfall, T.C. and Glowinski, J. (1978). Effects of acetylcholine and atropine on the release of ^{14}C-acetylcholine formed from U-^{14}C-glucose in rat brain cortical and striatal prisms. Naunyn-Schmiedeberg's Arch. Pharmacol., 303, 279-285.

Lippa, A.S., Critchett, D.J. and Bartus, R.T. (1979). Electrophysiological and biochemical evidence for age-related alterations in hippocampal cholinergic neurones. Soc. Neurosci. Abstr. 5, 11.

McGeer, E.G., Fibiger, H.C., McGeer, P.L. and Wickson, V. (1971). Aging and brain enzymes. Exp. Gerontol., 6, 391-396.

McGeer, P.L. and McGeer, E.G. (1976). Enzymes associated with the metabolism of catecholamines, acetylcholine and GABA in human controls and patients with Parkinson's disease and Huntington chorea. Neurochem. 26, 65-76.

Makman, M.H., Ahn, H.S., Thal, L.J., Sharpless, N.S., Dvorkin, B., Horawitz, S.G. and Rosenfeld, M. (1979). Aging and monoamine receptors in brain. Fed. Proc. 38, 1922-1926.

Mann, D.M.A. and Yates, P.O. (1979). The effect of aging on the pigmented nerve cells of the human locus caeruleus and substantia nigra. Acta. Neuropath. (Berl.), 47, 93-97.

Marchbanks, R.M. and Wonnacott, S. (1979). Relationship of choline uptake to acetylcholine synthesis and release. In Progress in Brain Research, vol. 49: The cholinergic synapse. (ed. S. Tucek), Elsevier, Amsterdam, p.77-88.

Meek, J.L., Bertilsson, L., Cheney, D.L., Zsilla, G. and Costa, E. (1977). Aging induced changes in acetylcholine and serotonin content in discrete brain nuclei. J. Gerontol., 12, 129-131.

Mehraein, P., Yamada, M. and Tarnowska-Dziduszko, E. (1975). Quantitative study on dendrites and dendritic spines in Alzheimer's disease and senile dementia. In Advances in Neurology, 12 (ed. G.W. Krentzberg), Raven Press, New York, pp. 453-458.

Melamed, E., Lavy, S., Siew, E., Bentin, S. and Cooper, G. (1978). Correlation between regional cerebral blood flow and brain atrophy in dementia. J. Neurol. Neurosurg. Psychiat., 41, 894-899.

Mirjakawa, T., Sumiyashi, S., Murayama, E. (1974). Capillary plaque-like degeneration in senile dementia. Acta Neuroptah. (Berl.), 29, 229-236.

265

Miller, A.K.H., Alston, R.C. and Corsellis, J.A.N. (19⁻:). Variations with age in the volumes of grey and white matter in the cerebral hemispheres of man: Measurements with an Image Analyser. Neuropath. Appl. Neurobiol., in press.

Morin, A.M. and Wasterlain, C.G. (1980). Aging and rat brain. Muscarinic receptors as measured by quinuclidinyl benzilate binding. Neurochem. Res., 5, 301-308.

Nikaido, T. Austin, J., Rinhart, R., Trueb, L., Hutchinson, J., Stukenbrok, H. and Miles, B. (1971). Studies in ageing of the brain. I. Isolation and preliminary characterisation of Alzheimer plaques and cores. Arch. Neurol. 25, 198-211.

Parkes, J.D., Marsden, C.D., Rees, J.E., Curzon, G., Kantamaneni, B.D., Knill-Jones, R., Akbar, A., Das, S. and Kataria, M. (1974). Parkinson's disease, cerebral arteriosclerosis and senile dementia. Clinical features and response to levodopa. Quart. J. Med. XLIII, 49-61.

Perl, D.P. and Brody, A.R. (1980). Alzheimer's disease: x-ray spectrometric evidence of aluminium accumulation in neurofibrillary tangle-bearing neurons. Science, 208, 297-299.

Perry, E.K., Gibson, P.H., Blessed, G., Perry, R.H. and Tomlinson, B.E. (1977a). Neurotransmitter enzyme abnormalities in senile dementia - choline acetyltransferase and glutamic acid decarboxylase activities in necropsy brain tissue. J. Neurol. Sci., 34, 247-265.

Perry, E.K., Perry, R.H., Gibson, P.H., Blessed, G. and Tomlinson, B.E. (1977b). A cholinergic connection between normal aging and senile dementia in the human hippocampus. Neurosci. Letters, 6, 85-89.

Perry, E.K., Tomlinson, B.E., Blessed, G., Bergmann, K., Gibson, P.H. and Perry, R.H. (1978). Correlation of cholinergic abnormalities with senile plaques and mental test scores in senile dementia. Brit. Med. J., 2, 1457-1459.

Plum, F. (1978). Metabolic dementias. In Alzheimer's Disease: Senile dementia and Related Disorders (eds. R. Katzman, R.D. Terry and K.L. Bick), Raven Press, New York, p.135-139.

Raiche, M.E., Grubb, R.L., Gado, M.H., Eichling, J.O. and Hughes, C.P. (1978). Cerebral hemodynamics and metabolism in dementia: Features distinguishing normal pressure hydrocephalus from atrophy. In Senile dementia: A biomedical approach (ed. K. Nandy), Elsevier/North Holland, New York. p.131-138.

Reisine, T.D., Yamamura, H.I., Bird, E.D., Spokes, E. and Enna, S.J. (1978). Pre- and post-synaptic neurochemical alterations in Alzheimer's disease. Brain Res.,159, 477-482.

Scheibel, A.B. and Tomiyasu, U. (1978). Dendritic Sprouting in Alzheimer's pre-senile dementia. Exp. Neurol. 60, 1-8.

Sims, N.R., Bowen, D.M., Smith, C.C.T., Flack, R.H.A., Davison, A.N., Snowden, J.S. and Neary, D. (1980). Glucose metabolism and acetylcholine synthesis in relation to neuronal activity in Alzheimer's disease. Lancet i, 333-336.

Sokoloff, L. (1976). Circulation and energy metabolism of the brain. In Basic Neurochemistry (eds. G.J. Siegel, R.W. Albers, R. Katzman and B.W. Agranoff), Little, Brown and Co., Boston, p.388-413.

Spillane, J.A., White, P., Goodhardt, M.J., Flack, R.H.A., Bowen, D.M. and Davison, A.N. (1977). Selective vulnerability of neurones in organic dementia. Nature (London), 266, 558-559.

Spokes, G.S. (1979). An analysis of factors influencing measurements of dopamine, noradrenaline, glutamate decarboxylast and choline acetylase in human post-mortem brain tissues. Brain, 102, 333-346.

Strong, R., Hsu, L., Hicks, P. and Enna, S.J. (1979). Age-related decrease in mouse brain cholinergic muscarinic receptor binding. Soc. of Neuroscience Abstracts, 5, 11.

Suzuki, K., Korey, S.R. and Terry, R.D. (1964). Studies on protein synthesis in brain microsome system. J. Neurochem., 11, 403-412.

Suzuki, K., Katzman, R. and Korey, S.R. (1965). Chemical studies on Alzheimer's disease. J. Neuropath. Exp. Neurol. 24, 211-224.

Terry, R.D. (1979). Aging of the central nervous system and senile dementia. In Muscle, Nerve and Brain Degeneration (eds. A.D. Kidman and J.K. Tomkins), Excerpta Medica, Amsterdam, pp. 187-202.

Tomlinson, B.E. (1977). Morphological changes and dementia in old age. In Ageing and Dementia (eds. W.L. Smith and M. Kinsbourne), Spectrum Press, New York, pp. 25-56.

Whalley, J.L., Urbaniak, S.J., Dang, L., Peutherer, J.F. and Christie, J.E. (1980). Histocompatibility antigens and antibodies to viral and other antigens in presenile dementia. Acta Psych. Scand., 61, 1-7.

White, P., Hiley, C.R., Goodhardt, M.J., Carrasco, L., Keet, J.P., Williams, J.E.J. and Bowen, D.M. (1977). Neocortical cholinergic neurones in elderly people. Lancet i, 668-670.

Winblad, B., Adolfsson, R., Gottfries, C.G., Oreland, L. and Roos, E.B. (1978). Brain monoamines, monoamine metabolites and enzymes in physiological ageing and senile dementia. In Recent Developments in Mass Spectrometry in Biochemistry and Medicine (ed. A. Frigerio), Plenum Press, New York, 253-267.

Wisniewski, W.M., Narang, H.K. and Terry, R.D. (1976). Neurofibrillary tangles of paired helical filaments. J. Neurol. Sci., 27, 173-181.

Yates, F.O., Mann, D.M.A., Lincoln, J., Toper, S. and Stamp, J.E. (1980) Widespread neuronal deficiency of RNA in Alzheimer's disease. Abstracts of International Society for Neurochemistry Meeting, Florence.

Yates, C.M., Allison, Y., Simpson, J., Maloney, A.F.J. and Gordon, A. (1979). Dopamine in Alzheimer's Disease and Senile Dementia, Lancet ii, 851-852.

Section V
Pathological States
(C)Parkinson's Disease

Parkinson's Disease as a Model for Behavioural Studies in Man

W. BIRKMAYER* and P. RIEDERER†

* Evangelical Hospital, A-1010 Vienna, and
† Ludwig Boltzmann-Institute of Clinical Neurobiology,
  Lainz-Hospital, A-1130 Vienna, Austria

The progress of medical science depends on the connection and cooperation of clinical observation and basic research. A success story of this cooperation is found in the understanding and treatment of Parkinson's disease where observations in the clinic were correlated with findings of decreased dopamine (DA) concentrations in post mortem brain (Ehringer and Hornykiewicz, 1960). Animal studies first identified DA as a neurotransmitter, and further such studies were preliminary to clinical treatment with L-dopa (Birkmayer and Hornykiewicz, 1961; Barbeau et al.,1961).

Basic research has shown Parkinson's disease to exhibit a degeneration of the melanin-containing cells of the substantia nigra, and our studies have shown deficits of noradrenaline (NA) and serotonin (5-HT) as well as DA in various brain stem nuclei. It is well established that the balance between DA and acetylcholine is responsible for normal motor response: that parkinsonism exhibits akinesia or, with an overdosage of L-dopa, hyperkinesia when this balance is disturbed. We report here that the balance between DA and NA, or DA and 5-HT, is also important. This is illustrated by both biochemical data and clinical observation.

Besides the well-known motor disturbances in Parkinson's disease, clinical observation has identified autonomic disturbances which include salivation, sweating and high temperature, and abnormalities of

271

affect (depression). From what we know of the function
of DA in man, these symptoms appear not to be directly
connected with a loss of this neurotransmitter.
Presumably though, there is a connection if DAergic
degeneration is the primary biochemical disturbance
in the disease. In order to throw light on the possible
biochemical correlates with those other symptoms, we
have undertaken measurements of neurotransmitters and
metabolites in brain regions taken post-mortem from
parkinsonians, depressives and control patients
(Birkmayer et al., 1977, Birkmayer and Riederer,1980).

However, before discussing these results, there is
more that we can learn from the clinic. Patients with
a severe motor dysfunction can sometimes walk with a
near-normal gait when in a clinical examination or
when meeting an old girlfriend, for example. Similarly,
very minor stressful situations can often induce hyperkinesia and tremor. We have found propranolol to be a
useful medication in such cases; an approach which is
also effective in non-neurological cases of anxiety.
Such behaviour is similar to the agitated behaviour
induced by anxiety in man which is perhaps more notable
in psychiatric cases (raptus phenomena).

The close relationship between affective or emotional
stimulation and motor function is notable not only in
neuropsychiatric disease but also in normal life.
This is common knowledge; typical examples can be seen
in all disciplines of sport. The difference between
the parkinsonian patients and "normal" man is the
lowered "affect threshold". The above example is
presumably NA dependent, responding as it does to a
ß-receptor blocker.

Motor function can also be decreased by an emotional
stimulus. "Freezing" in parkinsonian patients is
well known and can occur with very minor anxieties.
We can compare this to the "playing possum" reflex in
animals: danger induces a dead-still response.
Extreme fear in man can have the same "rigid with
fright" effect.

All these examples discussed above are archetypal
patterns of behaviour. That these extremes are found
in Parkinsonism indicates that the controlling
threshold has been surpassed and the effect is
triggered. The condition for normally adjusted

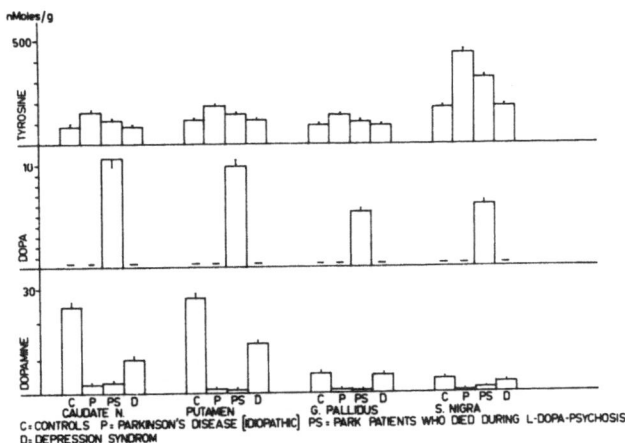

Figure 1   The concentration of tyrosine,dopa,
and dopamine in human brain areas
in different pathological stages

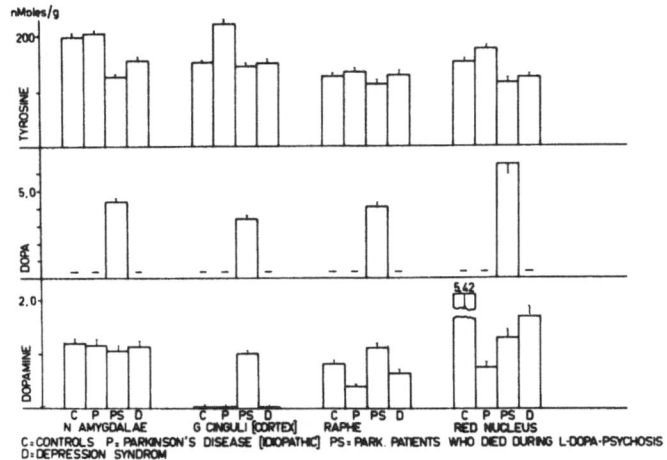

Figure 2   The concentration of tyrosine, dopa,
and dopamine in human brain areas
in different pathological stages

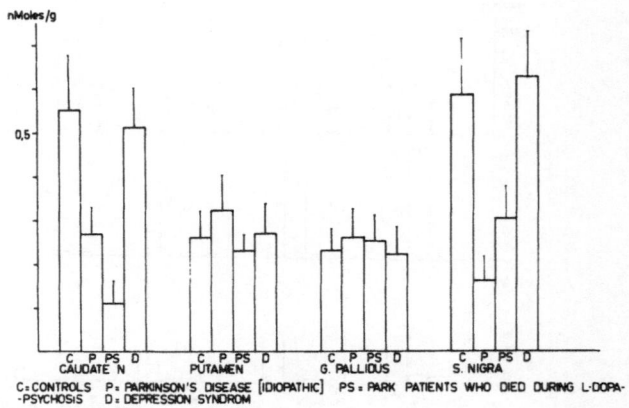

Figure 3   The concentration of noradrenaline
in human brain areas in different
pathological stages

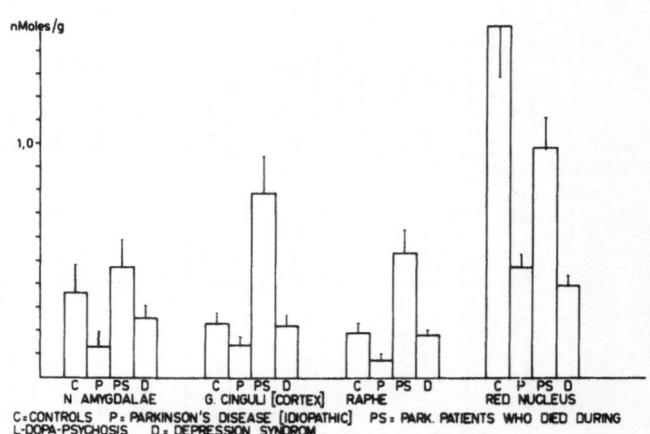

Figure 4   The concentration of noradrenaline
in human brain areas in different
pathological stages

behaviour is a functional balance of neurotransmitters. This balance can only be maintained by various feed-back mechanism. Thus the normal compensatory feedback mechanisms which control the motor response to an affective stimulus are upset since the neurotransmitter balance is disturbed.

An example of this autonomic feedback is in the control of temperature in man. A hot environment triggers an enlargement of blood capillaries in the skin. This feedback mechanism is regulated by 5-HT. A deficit of 5-HT in parkinsonian patients inhibits this physiolo-gical thermostat with the clinical result of hyper-thermia during a hot summer. This can be corrected by administration of the 5-HT precursor, 5-hydroxy-tryptophan (5-HTP) (Birkmayer and Neumayer, 1963).

Having discussed psycho-motor and autonomic disturban-ces, we come to some psychiatric abnormalities of Parkinsonism. It is commonly found that depressive phase can occur some years before the onset of the disease as well as during its course. There are also reports of L-dopa medication triggering depression. These apparently inconsistent reports can be explained by postulating a nuclei-specific imbalance of neuro-transmitters (Riederer and Birkmayer, 1980). Figures 1 - 6 indicate monoamine neurotransmitter levels in various brain regions in deceased depressed patients. We suggest that the lowered striatal DA is responsible for loss of drive, and that loss of 5-HT in the mid-brain (reticular formation) is responsible for sleeplessness. The high NA level in the reticular formation may correlate with anxiety. While no general disturbances in NA or 5-HT systems, reflected in body fluids is reported (Schildkraut, 1966; Coppen et al.,1967) there are various abnormalities in transmitter concen-trations in various brain regions (Birkmayer et al., 1977). Thus neither an administration of L-dopa or 5-HTP alone can really be effective. We have found that most of our parkinsonian patients respond to tricyclic antidepressant therapy. This supplement to L-dopa does not help the akinesia but the autonomic and affective decompensations which occur during the course of the disease can be substantially diminished.

A further example of the loss of neurotransmitter balance is in the occurrence of toxic delirium. Our results show (figs. 1 - 7) that in end-stage

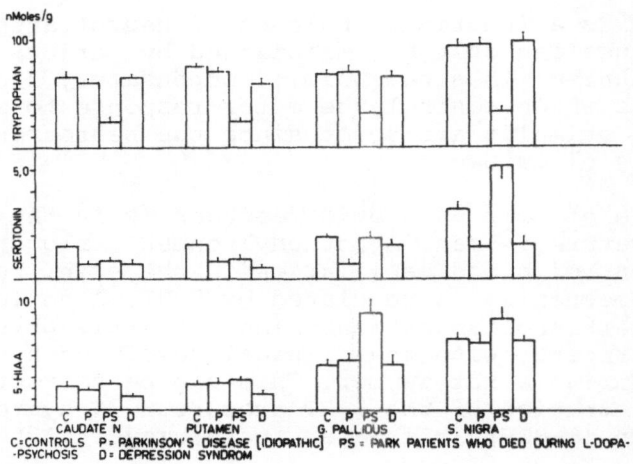

Figure 5   The concentration of tryptophan, serotonin, and 5-hydroxy-indole acetic acid in human brain areas in different pathological stages

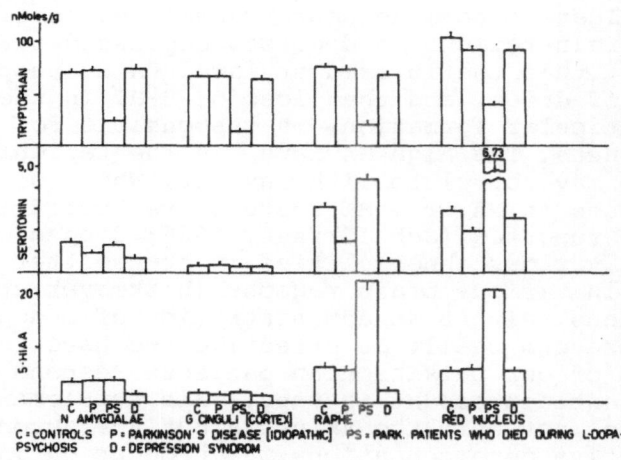

Figure 6   The concentration of tryptophan, serotonin, and 5-hydroxy-indole acetic acid in human brain areas in different pathological stages

psychosis a general disturbance, particularly in extrastriatal regions, is apparent in DA, NA and 5-HT. However, less severe cases often respond to a 5-HT precursor whereas with increasing L-dopa the psychosis is aggrevated. After injections of 5-HTP hallucinations, delusion and confusion disappear (Birkmayer and Neumayer 1972). Thus the 5-HT system might be counteracting the underlying system(s) responsible for the psychotic process. This response to a 5-HT precursor is only possible in the initial stages. In severe cases of Parkinson's disease the degeneration of DA neurons is so advanced that such medication cannot reestablish a functionally normal balance of transmitters.

We have identified two types of Parkinson's disease, malignant or benign, where onset of toxic delirium occurs, on average, 2.6 or 4.4 years respectively after L-dopa therapy is started (Birkmayer et al.,1979). The malignant cases presumably exhibit faster degeneration of DA neurons, eventually reaching a point beyond which no amount of L-dopa can provide adequate neuronal DA, and secondary 5-HT precursor therapy cannot overcome this primary problem.

Summarizing we conclude that a degeneration of specific neuronal regions in Parkinson's disease leads to a loss of balance between various biochemical transmitters. This loss of balance is responsible for the appearance of motor, psychomotor, autonomic and affective abnormalities which in normal man are prevented by regulatory feedback mechanisms. In parkinsonians these control mechanisms are diminished or lost due to the lower levels of neurotransmitters. Their substitution by specific precursors improves the clinical picture for as long as there are neurons available to synthesize and store the product neurotransmitters.

The Copernican era in biochemical brain research is now passed and it will soon be possible to restore the biochemical balance in human brain by the use of more highly specific neuropsychiatric drugs.

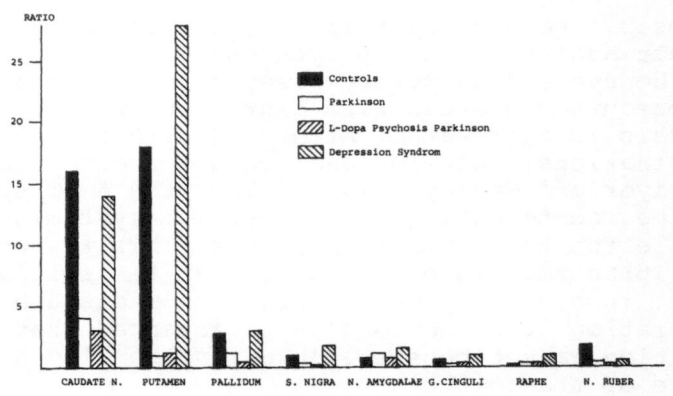

Figure 7   Dopamine/serotonin ratio in various
areas of the human brain

REFERENCES

Barbeau, A., Murphy, G.F., Sourkes T.L. (1961)
Excretion of Dopamine in Diseases of Basal Ganglia
Science 133, 1706

Birkmayer, W., Hornykiewicz, O.   (1961)   Der L-3,4-
Dioxyphenylalanin (=Dopa)-Effekt bei der Parkinson-
Akinese. Wien.klin.Wschr. 73, 787

Birkmayer, W., Neumayer, E., (1972)   Die Behandlung
der Dopa-Psychosen mit L-Tryptophan.   Nervenarzt 43,76

Birkmayer, W., Neumayer, E. (1963). Die Wärmeregulation beim
postencephalitischen Parkinsonismus. Nervenarzt 34, 373.

Birkmayer, W., Jellinger, K., Riederer, P. (1977)
Striatal and Extrastriatal Dopaminergic Functions.
In Psychobiology of the Striatum (eds. A.R.Cools,
A.H.M.Lohman and J.H.L. van den Bercken) North-
Holland, Amsterdam-New York-Oxford; pp 141

Birkmayer, W., Riederer, P., Youdim M.B.H. (1979)
Distinction between Benign and Malignant Type of
Parkinson's Disease. Clin.Neurol.Neurosurg. 81, 158

Birkmayer, W., Riederer P. (1980) Die Parkinson-
Krankheit. Springer Verlag Wien - New York

Coppen, A., Shaw, D.M., Herzberg, B., Maggs R. (1967)
Tryptophan in the Treatment of Depression. Lancet ii,
1178

Ehringer, H., Hornykiewicz, O., (1960) Verteilung
von Noradrenalin und Dopamin (3-Hydroxytyramin) im
Gehirn des Menschen und ihr Verhalten der Erkrankungen
des extrapyramidalen Systems. Klin.Wschr.38, 1236

Riederer, P., Birkmayer, W. (1980) A New Concept:
Brain Area Specific Imbalance of Neurotransmitters in
Depression Syndrome - Human Brain Studies.
In Enzymes and Neurotransmitters in Mental Disease
(eds. E.Usdin, T.L.Sourkes, and M.B.H.Youdim)
John Wiley, New York

Schildkraut, J.J., Green, R., Gordon, E.K., Durell, J.
(1966) Normetanephrine Excretion and Affective State
in Depressed Patients Treated with Imipramine.
Amer.J.Psych. 123, 690

Biopterin in the Brain of Controls and Patients with Parkinson's Disease and Related Striatal Degenerative Diseases: Application of New Biopterin Radioimmuno-assay

TOSHIHARU NAGATSU*, TOKIO YAMAGUCHI*, TAKESHI KATO*,
TAKASHI SUGIMOTOt, SADAO MATSUURAt, MIKI AKINO‡,
IKUKO NAGATSU§; REIJI IIZUKA¶ , and HIROTARO NARABAYASHI**

* Laboratory of Cell Physiology, Department of Life Chemistry,
  Graduate School at Nagatsuta, Tokyo Institute of Technology, Yokohama

† Department of Chemistry, Nagoya University, Nagoya

‡ Department of Biology, Tokyo Metropolitan University, Tokyo

§ Department of Anatomy, School of Medicine, Fujita—Gakuen University, Toyoake

¶ Department of Psychiatry and **Department of Neurology,
  School of Medicine, Juntends University, Tokyo, Japan

INTRODUCTION

We reported that tyrosine hydroxylase activity in the dopaminergic regions of nigro-striato-pallidal complex of the brain is greatly reduced (5-10% of controls) in all 12 cases of parkinsonian patients (Nagatsu, et al., 1977, 1979). The reduction in tyrosine hydroxylase activity in the parkinsonian brain was also reported by several groups (Lloyd, et al., 1975; McGeer and McGeer, 1976; Riederer, et al., 1979).

Tyrosine hydroxylase is a pterin-requiring monooxyge-nase (Nagatsu, et al., 1964), and L-erythro-tetrahydro-biopterin may be the natural cofactor (Brenneman and Kaufman, 1964). We found that concentration of the pterin cofactor, L-erythro-tetrahydrobiopterin, has an important role in the kinetic regulation of tyrosine hydroxylase (Numata, et al., 1977, Nagatsu, et. al., 1978). Importance of the pterin cofactor for the in vivo regulation of tyrosine hydroxylase was also reported (Kettler, et al., 1974; Weiner, et al., 1978; Nagatsu, 1979).

We have therefore examined biopterin concentrations in the postmortem human brain (caudate nucleus) and in the urine from controls and parkinsonian patients. There has been published no report on biopterin concentrations in the human brain to our knowledge. Since we have recently developed a raioimmunoassay for biopterin (Nagatsu, et al., 1979), we applied this new method for the assay of biopterin in postmortem brain tissues and urine of the humans.

BIOPTERIN IN HUMAN BRAIN FROM CONTROLS AND PARKINSONIAN PATIENTS

The control human brains and parkinsonian brains were reported in our previous paper (Nagatsu, et al., 1977, 1979). The parkinsonian patients had not taken L-DOPA during the last 6 months before death. The ages and postmortem periods of controls were similar to those of parkinsonian patients. The brain tissues were frozen and stored at -80°C. The frozen brain tissues (10 mg) were homogenized with 4 volumes of 0.1 M phosphate buffer, pH 7.5, and the homogenate was used for the assay of total biopterin and tyrosine hydroxylase activity.

Total biopterin concentrations in the human brain were measured by a new radioimmunoassay (Nagatsu, et al., 1979). In principle, specific antiserum for biopterin was prepared by using L-erythro-biopterinyl-caproyl-bovine serum albumin as antigen in rabbits, and L-erythro-biopterinyl-caproyl-[125I]tyramide was used as a radio-labelled ligand. The sensitivity of this radioimmunoassay has been further improved by using a new labelled ligand, a conjugate of L-erythro-biopterin with [125I]tyramine. Biopterin in the brain tissue is present mostly as the reduced forms, i.e. tetrahydro-biopterin or quinonoid dihydrobiopterin, and the reduced forms were first oxidized with iodine in acid to biopterin in order to measure the total biopterin, and then the biopterin was isolated on a small Dowex-50-H$^+$ column and assayed by the radioimmunoassay. We showed that the total biopterin concentration obtained by the radioimmunoassay agreed with the values obtained by Crithidia bioassay and high-performance liquid chromatography. In contrast, the literature values of tissue biopterin observed using the enzyme radioassay with phenylalanine hydroxylase are much higher than those obtained by our radioimmunoassay, Crithidia bioassay or high-performance liquid chromatography (Fukushima and Nixon, 1980).

Tyrosine hydroxylase activity in the brain homogenate was determined both by radioassay using L-[U-^{14}C]-tyrosine as substrate (Nagatsu, et al. , 1964; Nagatsu, 1973) or by a new high-performance liquid chromatography-voltammetry using L-tyrosine as substrate (Nagatsu, et al., 1979).

The reduced forms of biopterin (tetrahydrobiopterin and quinonoid dihydrobiopterin) are unstable and easily oxidized to 7,8-dihydrobiopterin, but the total biopterin concentrations in human and rat brains were found to be fairly stable. As shown in Figure 1, about 90 % of the zero time level of total biopterin in the human caudate nucleus remained at 3 hours at 20°C.

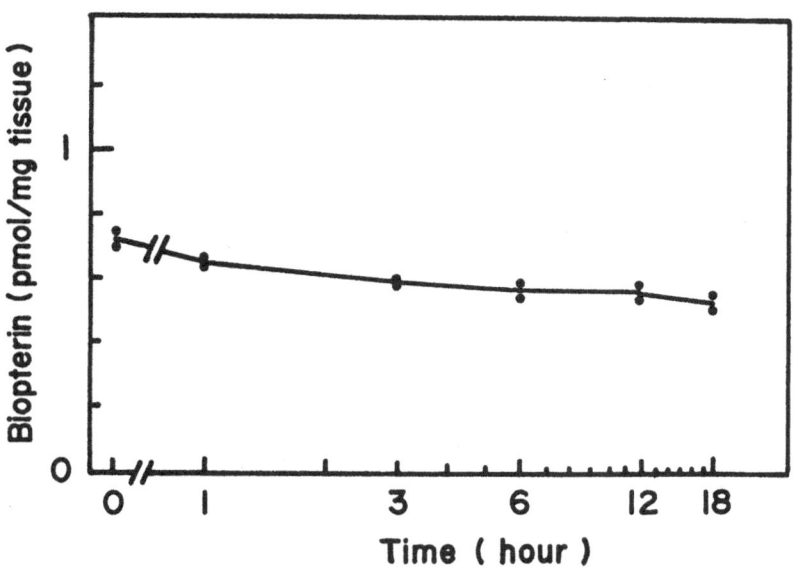

Figure 1. Postmortem changes of biopterin in human caudate nucleus. Brain tissue was left at room temperature (20°C) for the indicated time.

The total biopterin concentrations in the caudate nucleus of controls and parkinsonism are shown in Figure 2. The biopterin levels in parkinsonian brains were significantly lower than those of controls. The value of total biopterin in the human caudate nucleus of controls and parkinsonian patients was 1.25 ± 0.25 (range, 0.59-2.45) and 0.30 ± 0.06 (range, 0.12-0.62) nmol/g wet weight, mean ± S.E.M., respectively. The values of each one case of Shy-Drager syndrome and striato-nigral degeneration and two cases of tardive dyskinesia were 0.11, 0.66, and 0.52 (mean) nmol/g wet weight, respectively. These values were also lower than the control values.

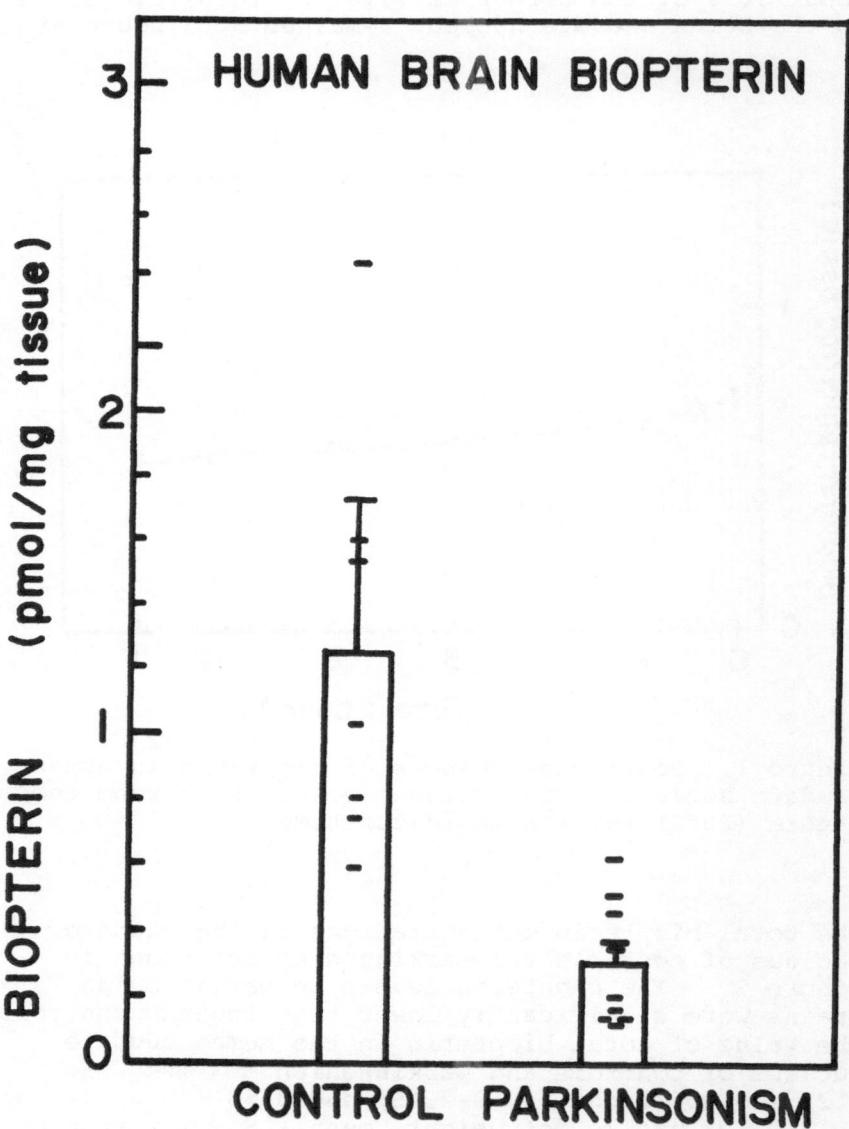

Figure 2. Biopterin concentrations in the caudate nucleus of controls and parkinsonian patients

Significant positive correlation between the total biopterin level and the activity of tyrosine hydroxylase in the caudate nucleus of controls and parkinsonian cases was observed; r=0.87 (p < 0.01). Such significant correlation between tyrosine hydroxylase activity and tetrahydrobiopterin was also found in brain regions of rats (Bullard et al., 1978; Levine, et al., 1979). The decrease of tyrosine hydroxylase (to 5-10 % of the controls) was more marked than the decrease of biopterin (to about 25 % of the controls).

These results confirmed that the concentrations of biopterin, an essential cofactor for tyrosine hydroxylase, were significantly reduced in the striatum of parkinsonian patients as compared with those of age-matched controls. We had previously reported a great reduction in Vmax of tyrosine hydroxylase activity under saturating pterin cofactor concentrations in the nigro-striatal dopaminergic tissues from parkinsonian patients (Nagatsu, et al., 1977, 1979). This indicates that the protein concentration of tyrosine hydroxylase is decreased. Since both tyrosine hydroxylase and biopterin are localized in the same dopaminergic cells, simultaneous reduction of both the enzyme and the cofactor due to cell loss of dopaminergic neurons can be expected. Greater reduction of tyrosine hydroxylase activity as compared with the reduction of biopterin in parkinsonian brain can be explained by the fact that biopterin may be rich in the dopamine cells which decrease in Parkinson's disease, but may also exist in glial cells which do not contain tyrosine hydroxylase.

Decrease in biopterin in parkinsonian brain suggests a general disturbance in the biosynthesis of monoamines (catecholamines and indoleamines), because both tyrosine hydroxylase and tryptophan hydroxylase require biopterin as cofactor. Therefore, the decrease in biopterin may account for the reduction of not only dopamine, but also noradrenaline, adrenaline, and serotonin in parkinsonian brain.

In accordance with our results on reduction of biopterin in the parkinsonian brain, tetrahydrobiopterin in the cerebrospinal fluid was found to be lower in parkinsonian patients than that in the control patients (Lovenberg, et al., 1979). Our results could suggest that the tetrahydrobiopterin in the cerebrospinal fluid may be derived from the brain tissues.

In order to see whether or not the reduction in biopterin in parkinsonian brain is specific for the brain,

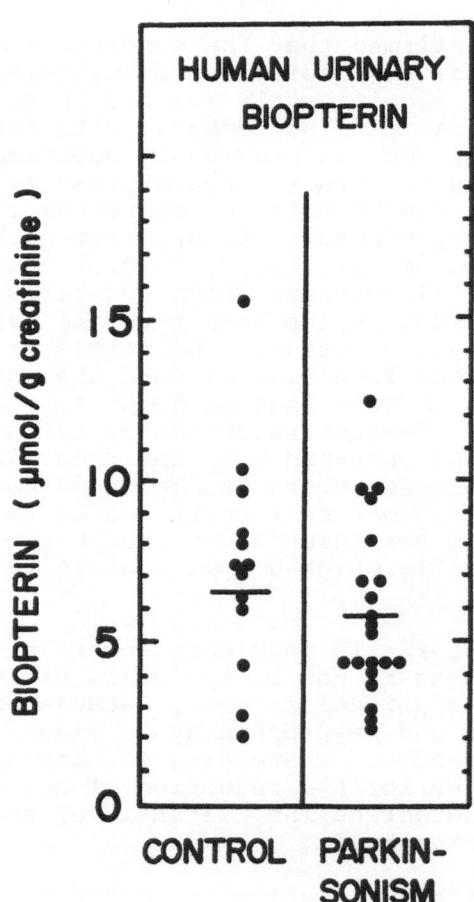

Figure 3.   Biopterin concentrations in the urine of
13 normal subjects and of 21 parkinsonian patients.
Mean ages of the subjects were 44.2 ± 6.2 years for
the normal controls and 60.5 ± 3.0 years for the
parkinsonian patients.

we have also measured biopterin in the urine from 13 normal subjects and from 21 parkinsonian patients. As shown in Figure 3, total urinary biopterin levels in parkinsonian patients [5.84 ± 0.59 (S.E.M.) μmol/g creatinine] were slightly lower than those of normal controls [6.53 ± 0.68 (S.E.M.) μmol/g creatinine], but the difference is not statistically significant. Biopterin may be synthesized in the human kidney and excreted into the urine. Therefore, the synthesis of biopterin in the kidney may not be significantly different between normal controls and parkinsonian patients. This may be in agreement with the concept of specific degeneration in the nigro-striatal dopaminergic neurons in parkinson's disease.

CONCLUSION

Biopterin in the human brain from controls and parkinsonian patients was measured by our newly established radioimmunoassay procedure. Both tyrosine hydroxylase and biopterin, the cofactor of the enzyme, was found to be greatly decreased in parkinsonian brain as compared with normal controls. In contrast, urinary excretion of biopterin in parkinsonian patients was lower than, but not significantly different from, that of normal subjects. These results suggest decreased synthesis rate of biopterin in the parkinsonian brain.

An interesting question is whether the reduction of biopterin in the dopaminergic neurons of parkinsonian patients proceeds the decrease of tyrosine hydroxylase and could have some causal relation to the cell loss, and whether or not administration of biopterin or tetrahydrobiopterin could prevent the disease process of parkinson's disease. Narabayashi et al. are currently planning such clinical studies.

ACKNOWLEDGMENT

We wish to thank Mr. Kazuhiro Oka for the assay of tyrosine hydroxylase in a part of brain samples. This study is supported in part by the Research Grant for the Intractable Diseases from the Ministry of Health and Welfare of Japan and by the Research Grant from Takeda Science Foundation,Osaka, Japan, to T.N., which are gratefully acknowledged.

REFERENCES

Brenneman, A. R. and Kaufman, S. (1964). The role of tetrahydropteridines in the enzymatic conversion of tyrosine to 3,4-dihydroxyphenylalanine. Biochem. Biophys. Res.Commun., 17, 177-183.

Bullard, W. P., Guthrie, P. B., Russo, P. B. and Mandell, A. J. (1978). Regional and subcellular distribution and some factors in the regulation of reduced pterins in rat brain. J. Pharmacol. Exp. Therap., 206, 4-20

Fukushima, T. and Nixon, J. C. (1980). Analysis of reduced forms of biopterin in biological tissues and fluids. Anal. Biochem., 102, 176-188.

Kettler, R., Bartholini, G. and Pletscher, A. (1974). In vivo enhancement of tyrosine hydroxylation in rat striatum by tetrahydrobiopterin. Nature, 249, 476-478.

Levine, R. A., Kuhn, D. M. and Lovenberg, W. (1979). The regional distribution of hydroxylase cofactor in rat brain. J. Neurochem., 32, 1575-1578.

Lloyd, K. G., Davidson, L. and Hornykiewicz, O. (1975). The neurochemistry of Parkinson's disease: effect of L-DOPA therapy. J. Pharmacol. Exp. Therap., 195, 453-464.

Lovenberg, W., Levine, R. A., Robinson, D. S., Ebert, M., Williams, A. C. and Calne, D. B. (1979). Hydroxylase cofactor activity in cerebrospinal fluid of normal subjects and patients with Parkinson's disease. Science, 204, 624-626.

McGeer, P. L. and McGeer, E. G. (1976). Enzymes associated with the metabolism of catecholamines, acetylcholine and GABA in human controls and patients with Parkinson's disease and Huntington's chorea. J. Neurochem., 6, 65-76.

Nagatsu, T. (1973). Biochemistry of Catecholamines, Univ. Tokyo Press and Univ. Park Press, Tokyo and Baltimore.

Nagatsu, T. (1979) Regulation of tyrosine hydroxylase. In Frontiers in Catecholamine Research, (eds. E. Usdin and S. Snyder), Pergamon Press, Oxford.

Nagatsu, T., Kato, T., Numata(Sudo), Y., Ikuta, K., Sano, M., Nagatsu, I., Kondo, Y., Inagaki, S., Iizuka, R., Hori, A. and Narabayashi, H. (1977). Phenyl-ethanolamine-N-methyltransferase and other enzymes of catecholamine metabolism in human brain. Clin. Chim. Acta, 75, 221-232.

Nagatsu, T., Kato, T., Nagatsu, I., Kondo, Y., Inagaki, S., Iizukz, R. and Narabayashi, H., (1979). Catechol-amine-related enzymes in the brain of patients with parkinsonism and Wilson's disease. Adv. in Neurol., 24, 283-292.

Nagatsu, T., Levitt, M and Udenfriend, S. (1964). Tyrosine hydroxylase. The initial step in norepine-phrine biosynthesis. J. Biol. Chem., 239, 2910-2917.

Nagatsu, T., Numata(Sudo), Y., Kato, T., Sugiyama, K. and Akino, M. (1978). Effects of melanin on tyrosine hydroxyalse and phenylalanine hydroxylase. Biochim. Biophsy. Acta, 523, 47-52.

Nagatsu, T., Oka, K. and Kato, T. (1979). Highly sen-sitive assay for tyrosine hydroxylase activity by high-performance liquid chromatography. J. Chromatogr. 163, 247-252.

Nagatsu, T., Yamaguchi, T., Kato, T., Sugimoto, T., Matsuura, S., Kobayashi, K., Akino, M., Tsushima, S., Nakazawa, N. and Ogawa, H. (1979) Proc. Japan Acad., 55, Ser. B, 317-322.

Numata(Sudo), Y., Kato, T., Nagatsu, T., Sugimoto, T. and Matsuura, S. (1977). Effects of stereochemical structures of tetrahydrobiopterin on tyrosine hydroxy-lase. Biochim. Biophys. Acta, 480, 104-112.

Riederer, P., Rausch, W. -D., Birkmayer, W., Jellinger, K. and Seemann, D. (1978). CNS modulation of adrenal tyrosine hydroxylase in Parkinson's disease and metabolic encephalopathies. J. Neural Transmission, Suppl. 14, 121-131.

Weiner, N., Lee, F. -L., Dreyer, E. and Barnes, E. (1978). The activation of tyrosine hydroxylase in noradrenergic neurons during acute nerve stimulation. Life Sci., 22, 1197-1216.

Catecholaminergic Enzymes in Parkinson's Disease and Related Extrapyramidal Diseases

TOSHIHARU NAGATSU*, KAZUHIRO OKA*, TOSHIFUMI YAMAMOTO*, HIROAKI MATUSUI†, TAKESHI KATO*, CHOSABURO YAMAMOTO†, IKUKO NAGATSU‡, REIJI IIZUKA§ AND HIROTARO NARABAYASHI¶.

* Laboratory of Cell Physiology, Department of Life Chemistry, Graduate School of Nagatsuta, Tokyo Institute of Technology, Yokohama
† Department of Physiology, Faculty of Medicine, Kanazawa University, Kanazawa
‡ Department of Anatomy, School of Medicine, Fujita-Gakuen University, Tayoake
§ Department of Psychiatry and ¶ Department of Neurology, School of Medicine, Juntendo University, Tokyo, Japan

The observation on the decrease in dopamine(Ehringer and Hornykiewicz, 1960) and in aromatic L-amino acid decarboxylase activity with DOPA as a substrate (DOPA decarboxylase activity)(Lloyd and Hornykiewicz, 1970) in the dopaminergic regions of the parkinsonian brain indicated the disturbance in dopamine biosynthesis.

Since 1974 we have been studying changes in the activity of catecholamine-related enzymes in the human brain from normal controls and from parkinsonian patients (Nagatsu, et al., 1977; 1979a).

We have examined 12 cases of parkinsonian patients who had not taken L-DOPA during the last 6 months before death. Each one case of striato-nigral degeneration, Shy-Drager syndrome, Wilson's disease, and Huntington's chorea, and two cases of tardive dyskinesia have also been examined. Data on the patients examined are shown in Table 1. The control human brains and parkinsonian brains from cases 1 to 8 were reported in our previous paper (Nagatsu, et al., 1977; 1979a).

Mean values of ages and postmortem period of the 11 controls and 12 parkinsonian patients were 54.2 ± 5.4 years and 68.6 ± 2.8 years, and 5.8 ± 1.2 hours and 5.0 ± 1.2 hours, respectively. The activity of tyrosine hydroxylase or DOPA decarboxylase was measured both by the previously reported radiochemical method with L-[U-^{14}C]tyrosine or DL-[1-^{14}C]DOPA as substrate (Nagatsu, et al., 1977; 1979a) and by the new high-performance liquid chromatography (HPLC)-

TABLE 1. Case histories of patients with parkinsonism and related extrapyramidal diseases

Case	Sex/Age	Cause of death	Postmortem period (hour)
Parkinsonism			
1 S.M.	M/73	Pneumonitis	4
2 I.N.	F/63	Unknown	6
3 M.O.	F/58	Heart failure	2
4 O.T.	F/78	Heart failure	10
5 M.S.	M/63	Heart failure	2.5
6 S.H.	F/61	Heart failure	6
7 N.O.	M/72	Peritonitis	1.5
8 S.K.	F/82	Heart failure	1
9 B.N.	M/76	Heart failure	7
10 T.M.	M/49	Gastric bleeding	3
11 K.Y.	F/70	Pneumonia	2
12 Y.M.	F/78	Heart failure	15
Striato-nigral degeneration			
S.H.	M/74	Respiratory failure	1.5
Shy-Drager syndrom			
Y.I.	M/53	Pneumonia	4
Wilson's disease			
T.S.	M/22	Pulmonary embolism	3
Huntington's chorea			
A.K.	F/53	Subdural hygroma	2
Tardive dyskinesia			
T.M.	M/63	Myocardial infarction	9
M.U.	M/39	Myocardial infarction	3

voltammetry (using Yanaco L-2000 with VMD-100 voltammetry detector, Kyoto, Japan) with L-tyrosine or
L-DOPA as substrate (Nagatsu, et al., 1979b; 1979c).
Values found with the HPLC assays were comparable to
those of radiometric methods. Dopamine-β-hydroxylase
activity was determined spectrophotometrically (Kato, et al.,1978).

TYROSINE HYDROXYLASE ACTIVITY IN HUMAN BRAIN

The activity of tyrosine hydroxylase, in terms of
nmol/hour/g tissue or pmol/min/mg protein, is shown
in Table 2. Tyrosine hydroxylase activity in all 12
parkinsonian patients was markedly decreased (4-12 %
of controls) especially in the putamen, caudate
nucleus, pallidum, and substantia nigra. It was also
significantly reduced in the locus coeruleus.

Reduction in tyrosine hydroxylase activity in
parkinsonian brains has also been reported by several
groups (Lloyd, et al., 1975; McGeer and McGeer, 1976;
Riederer, et al., 1979).

Great reduction in tyrosine hydroxylase activity in
dopaminergic regions was found in striato-nigral
degeneration and Shy-Drager syndrome; the enzyme
activity was also moderately decreased in noradrenergic regions. In tardive dyskinesia, there was a
tendency of decrease in tyrosine hydroxylase especially in nerve terminals of dopaminergic regions.

DOPA DECARBOXYLASE ACTIVITY IN HUMAN BRAIN

DOPA decarboxylase activity was shown in Table 3.
In accordance with the first observation by Lloyd and
Hornykiewicz (1970), DOPA decarboxylase activity was
also decreased in 8 cases of parkinsonian patients in
the putamen, caudate nucleus, pallidum, and substantia
nigra. But the activity in 4 cases were similar to or
even higher than that in controls. Therefore, the
mean values of all parkinsonian brains were not significantly different from those of controls. This
supports our previous observation on the presence of
parkinsonism (about 1/3 of our cases) with low tyrosine hydroxylase and yet normal DOPA decarboxylase.
Histopathological study is under way on the morphological differences in the brain of parkinsonian
patients with low tyrosine hydroxylase and low DOPA
decarboxylase, and with low tyrosine hydroxylase and
normal DOPA decarboyxlase, respectively.

TABLE 2A. Tyrosine hydroxylase activity in brain regions from patients with parkinsonism and related extrapyramidal diseases

Brain region	Tyrosine hydroxylase activity (mean ±S.E.M.)						
	Controls	PD	SND	SD	WD	HC	TD
	(nmol/hour/g tissue)						
Caudate nucleus	52.3 ±11.7(7)	5.8 ±1.9(12)*	3.5	23.1	19.7	20.9	9.5(2)
Putamen	76.0 ±26.9(8)	2.8 ±1.1(12)*	0.0	0.0	18.8	23.1	7.6
Pallidum	49.5 ±11.7(6)	5.1 ±3.1(12)*	0.41	4.3	40.3	4.3	3.9
Substantia nigra	99.7 ±27.6(6)	12.1 ±5.1(12)*	0.0	2.5	210	17.1	29.0(2)
Locus coeruleus	154 ±25(4)	21.0 ±9.0(11)*	10.8	9.2	——	——	15.1(2)
Hypothalamus	13.4 ±3.4(7)	10.2 ±3.3(10)	0.0	3.9	17.6	2.4	0.0

Numbers of samples are given in parentheses.
*$p < 0.01$ for difference between controls and parkinsonism.

PD:Parkinsonism, SN:Striato-nigral degeneration, SD:Shy-Drager syndrome, WD:Wilson's disease, HC:Huntington's chorea, TD:Tardive dyskinesia.

TABLE 2B. Tyrosine hydroxylase activity in brain
regions from patients with parkinsonism and related
extrapyramidal diseases

Brain region	Tyrosine hydroxylase activity (mean±S.E.M.)						
	Controls	PD	SND	SD	WD	HC	TD
	(pmol/min/ mg protein)						
Caudate nucleus	10.2 ±2.8(7)	0.98 ±0.31(12)*	0.61	4.00	4.22	3.96	1.62(2)
Putamen	15.2 ±6.4(8)	0.39 ±0.15(12)*	0.00	0.00	4.24	4.11	0.77
Pallidum	8.92 ±2.27(6)	1.00 ±0.61(12)*	0.05	0.55	8.59	0.65	0.05
Substantia nigra	18.5 ±5.4(6)	2.17 ±0.64(12)*	0.00	0.50	51.9	3.06	5.65(2)
Locus coeruleus	36.0 ±6.3(4)	4.72 ±2.10(11)*	2.00	1.82	——	——	3.69(2)
Hypothalamus	2.82 ±0.67(7)	1.77 ±0.57(10)	0.00	0.57	3.37	0.49	0.00

Numbers of samples are given in parentheses.
*$p < 0.01$ for difference between controls and
parkinsonism.

PD:Parkinsonism, SN:Striato-nigral digeneration,
SD:Shy-Drager syndrome, SD:Wilson's disease,
HC:Huntington's chorea, TD:Tardive dyskinesia.

TABLE 3A.   DOPA decarboxylase activity in brain regions from patients with parkinsonism and related extra-pyramidal diseases

Brain region	DOPA decarboxylase activity (mean ± S.E.M.)						
	Controls	PD	SND	SD	WD	HC	TD
	(nmol/hour/g tissue)						
Caudate nucleus	42.0 ±22.0(8)	32.8 ±11.2(12)	9.1	0.0	96.3	193	25.2(2)
Putamen	55.9 ±36.1(6)	17.4 ±6.3(12)	0.0	2.0	89.0	125	——
Pallidum	25.4 ±21.9(6)	16.8 ±8.5(12)	——	0.1	160	13.1	——
Substantia nigra	13.5 ±10.7(6)	27.6 ±15.6(12)	15.5	0.0	699	105	264(2)
Locus coeruleus	38.6 ±25.0(5)	60.1 ±22.5(10)	7.0	0.0	——	——	377(2)
Hypothalamus	30.8 ±14.3(7)	47.9 ±20.5(10)	10.1	0.0	165	12.2	4.5

Numbers of samples are given in parentheses.
PD:Parkinsonism, SN:Striato-nigral degeneration,
SD:Shy-Drager syndrom, WD:Wilson's disease,
HC:Huntington's chorea, TD:Tardive dyskinesia.

TABLE 3B. DOPA decarboxylase activity in brain regions from patients with parkinsonism and related extra-pyramidal diseases

Brain region	DOPA decarboxylase activity (mean ± S.E.M.)						
	Controls	PD	SND	SD	WD	HC	TD
	(pmol/min/mg protein)						
Caudate nucleus	8.60 ±5.06(8)	6.10 ±2.21(12)	1.4	0.0	20.0	37.0	3.7(2)
Putamen	11.9 ±8.5(6)	3.10 ±1.15(12)	0.0	0.2	19.4	22.5	——
Pallidum	4.82 ±4.22(6)	3.57 ±1.78(12)	——	0.0	32.5	2.0	——
Substantia nigra	2.54 ±2.02(6)	4.59 ±2.52(12)	2.6	0.0	189	18.9	42.2(2)
Locus coeruleus	9.35 ±6.28(5)	14.0 ±5.4(10)	1.95	0.0	——	——	128(2)
Hypothalamus	6.77 ±3.22(7)	9.11 ±4.01(10)	2.9	2.0	31.1	2.5	0.8

Numbers of samples are given in parentheses.
PD:Parkinsonism, SN:Striato-nigral degeneration,
SD:Shy-Drager syndrome, WD:Wilson's disease,
HC:Huntington's chorea, TD:Tardive dyskinesia.

Great reduction in DOPA decarboxylase activity, especially in nigro-striatal dopaminergic regions, was
also found in striato-nigral degeneration and Shy-
Drager syndrome.

DOPAMINE-β-HYDROXYLASE ACTIVITY IN HUMAN BRAIN

Dopamine-β-hydroxylase activity (Table 4) was high in
the locus coeruleus and hypothalamus of the control
brains.   It was lower in parkinsonism, striato-nigral
degeneration and Shy-Drager syndrome than in the
controls, suggesting the presence of a moderate impairment in the noradrenergic neurons in parkinsonism
and related extrapyramidal diseases. Although copper
concentrations in the brain were found to be greatly
increased in a patient with Wilson's disease,the
activity of dopamine-β-hydroxylase, a copper enzyme,
was rather low in the hypothalamus (Nagatsu et al.,
1979a).

SUMMARY

Although further work on more cases is needed, the
present results may be tentatively summarized as
shown in Table 5.   Tyrosine hydroxylase activity in
the nigro-striatal dopaminergic neurons is greatly
decreased in parkinsonism, striato-nigral degeneration,
and Shy-Drager syndrome. DOPA decarboxylase activity
is also decreased in these striatal diseases, but
there may be some cases (4 from 12 cases) of parkinsonism with low tyrosine hydroxylase and normal DOPA
decarboxylase. Not only tyrosine hydroxylase and
DOPA decarboxylase, but also dopamine-β-hydroxylase
in locus coeruleus and hypothalamus are moderately
decreased in parkinsonism, striato-nigral degeneration
and Shy-Drager syndrome. This further confirms our
previous results (Nagatsu, et al., 1977; 1979a) and
indicates partial impairment of noradrenergic neurons
in parkinsonism and related striatal diseases.

ACKNOWLEDGMENTS

This study is supported in part by the Research
Grant for the Intractable Diseases from the Ministry
of Health and Welfare of Japan and by the Research
Grant from the Mitsubishi Foundation, Tokyo, Japan,
to T.N., which are gratefully acknowledged.

298

TABLE 4. Dopamine-β-hydroxylase activity in brain regions from patients with parkinsonism and related extrapyramidal diseases

Brain region	Dopamine-β-hydroxylase activity (mean±S.E.M.)						
	Controls	PD	SND	SD	WD	HC	TD
(nmol/hour/g tissue)							
Locus coeruleus	876(2)	109 ±41(4)	242	303	——	204	1223(2)
Hypothalamus	264 ±80(9)	94.4 ±22.6(10)*	113	14.4	51.4	248	146
Thalamus	41.4 ±13.8(7)	52.6 ±14.4(12)	63.1	29.7	19.0	223	150
(pmol/min/mg protein)							
Locus coeruleus	188 ±76(2)	23.3 ±8.4(4)	21.0	44.0	——	36.2	253
Hypothalamus	58.9 ±17.9(9)	16.8 ±4.2(9)*	38.9	12.3	9.6	37.7	29.3
Thalamus	6.00 ±2.36(7)	9.84 ±4.34(12)	8.75	6.28	3.5	34.5	21.7

Numbers of samples are given in parentheses.
*$p < 0.05$ for difference between controls and parkinsonism.
PD:Parkinsonism, SN:Striato-nigral digeneration, SD:Shy-Drager syndrome, WD:Wilson's disease, HC:Huntington's chorea, TD:Tardive dyskinesia.

TABLE 5. Probable canges in catecholamine-related enzymes in the brain of patients with Parkinson's disease and related extrapyramidal diseases

Diseases (No. of cases)	Nigro-striatal dopaminergic region		Locus coeruleus-hypothalamus noradreneric region		
	TH	DDC	TH	DDC	DBH
Parkinson's disease(12)	↓↓	↓↓↓ →	↓	↓	↓
Striato-nigral degeneration(1)	↓↓ a	↓↓	↓	↓	↓
Shy-Drager syndrome(1)	↓↓ a	↓↓	↓	↓	↓
Wilson's disease(1)	→	→	n.d.	n.d.	↓
Huntington's chorea(2)	→	→	n.d.	n.d.	→
Tardive dyskinesia(2)	↓ b	↓ b	↓ b	↓ b	→

A:especially at putamen,
b:especially at nerve terminals,
n.d.:not determined. ↓↓:greatly decreased,
↓ :decreased, and →:normal.

TH:tyrosine hydroxylase, DDC:DOPA decarboxylase, DBH:dopamine-β-hydroxylase.

REFERENCES

Ehringer, H. and Hornykiewicz, O. (1960). Verteilung von Noradrenalin und Dopamin (3-Hydroxytyramin) in Gehirn des Menschen und ihr Verhalten bei Erkrankungen des Extrapyramidalen Systems. Klin. Wochenschr., 38 1236-1239.

Kato, T., Wakui, Y., Nagatsu, T., and Ohnishi, T. (1978). An improved dual-wavelength spectrophotometric assay for dopamine-β-hydroxylase. Biochem. Pharmacol., 27, 829-831.

Lloyd, K. G., Davidson, L. and Hornykiewicz, O. (1975). The neurochemistry of Parkinson's disease: effect of L-DOPA therapy. J. Pharmacol. Exp. Therap., 195, 453-464.

Lloyd, K. and Hornykiewicz, O. (1970). Parkinson's disease: activity of L-DOPA decarboxylase in discrete brain regions. Science, 170, 1212-1213.

McGeer, P. L. and McGeer, E. G. (1976). Enzymes associated with the metabolism of catecholamines, acetylcholine and GABA in human controls and patients with Parkinson's disease and Huntington's chorea. J. Neurochem., 6, 65-76.

Nagatsu, T., Kato, T., Numata(Sudo), Y., Ikuta, K., Sano, M., Nagatsu, I., Kondo, Y., Inagaki, S., Iizuka, R., Hori, A., and Narabayashi, H. (1977). Phenylethanolamine-N-methyltransferase and other enzymes of catecholamine metabolism in human brain. Clin. Chim. Acta, 75, 221-232.

Nagatsu, T., Kato, T., Nagatsu, I., Kondo, Y., Inagaki, S., Iizuka, R., and Narabayashi, H. (1979a). Catecholamine-related enzymes in the brain of patients with parkinsonism and Wilson's disease. Adv. Neurol., 24, 283-292.

Nagatsu, T., Oka, K., and Kato, T. (1979b). Highly sensitive assay for tyrosine hydroxylase activity by high-performance luquid chromatography. J. Chromatogr., 163, 247-252.

Nagatsu, T., Yamamoto, T., and Kato, T. (1979c). A new and highly sensitive voltammetric assay for aromatic L-amino acid decarboxylase activity by high-performance liquid chormatogrpahy. Anal.

Biochem., 100 160-165.

Riederer, P., Rausch, W. -D., Birkmayer, W., Jellinger,
K., and Seemann, D. (1978). CNS modulation of
adrenal tyrosine hydroxylase in Parkinson's disease
and metabolic encephalopathies. J. Neural Transmission, Suppl., 14, 121-131.

Brain Dopamine Receptors in Parkinson's Disease: Involvement with Clinical Features and Therapeutic Responses

U. K. RINNE

Department of Neurology, University of Turku, Turku, Finland

It is now generally accepted that the loss of dopaminergic substantia nigra neurons and dopamine deficiency in the corpus striatum play an essential role in the pathophysiology of Parkinson's disease. This was well established by Ehringer and Hornykiewicz (1960) and has subsequently been confirmed by several studies showing a decrease of dopamine and its metabolite, homovanillic acid, in the brain (Bernheimer et al., 1973; Lloyd et al., 1973; Riederer and Wuketich, 1976; Rinne and Sonninen, 1973; Rinne et al., 1971, 1974). Besides the loss of dopamine neurons, there is recent evidence of alterations in the striatal dopamine receptors in Parkinson's disease (Reisine et al., 1977; Lee et al., 1978; Rinne et al., 1979).

The present paper is concerned with further studies of dopamine receptors in the parkinsonian brain. We have attempted to obtain further clarification about the involvement of brain dopamine receptors with various clinical variables and therapeutic responses of the patients.

RESULTS

1. Striatal Dopamine Receptors in Control and Parkinsonian Patients

Striatal dopamine receptors were studied in post-mortem brain samples of 44 patients with Parkinson's disease by the radioligand-binding technique using ^{3}H-spiroperidol.

In controls the highest number of binding sites for ^{3}H-spiroperidol in the striatum was found in the caudate nucleus and putamen compared with that in the pallidum (Table 1). Treatment with neuroleptic drugs increased significantly (P< 0.001) the specific binding of ^{3}H-spiroperidol in all these striatal nuclei of control patients without extrapyramidal disorders. Scatchard analysis showed that with neuroleptics the number of receptors in these control patients was significantly (P< 0.001) increased. But there was no significant change in the mean value of the dissociation constant, although there was a tendency towards increased affinity and in four patients who had long-term treatment with neuroleptics the affinity was clearly increased.

The specific binding of ^{3}H-spiroperidol was significantly reduced in the caudate nucleus (P< 0.001) and putamen (P< 0.001) of a group of five parkinsonian patients who had not received any levodopa therapy (Table 1). Scatchard analysis showed that there was a significant increase (P <0.01) in the receptor number (Table 1), but no significant change in the dissociation constant. On the other hand, in the pallidum of these parkinsonian patients there was no significant change in ^{3}H-spiroperidol binding (Table 1).

Most of the parkinsonian patients, however, showed a significantly increased binding of ^{3}H-spiroperidol in the caudate nucleus and putamen compared to the former patients or controls. Scatchard analysis showed that in these patients the number of dopamine receptors was significantly increased (Table 1). Moreover, in four of these patients the affinity was also increased, although the mean value of the dissociation constant did not differ significantly from that of the controls. There were no significant differences in the clinical variables between these groups of parkinsonian patients, neither were there significant changes in the deep-frozen storage time of the brain samples. However, our patients were relatively severely disabled at the time of death and therefore the greatest caution should be observed in drawing valid conclusions from the relationship between the changes in the receptors and clinical variables.

Table 1 shows also that parkinsonian patients suffering from psychotic episodes and treated with neuroleptic drugs before death also had a significantly higher binding of ^{3}H-spiroperidol than controls, not only in the caudate nucleus and putamen but also in the pallidum. Scatchard analysis showed an increased number of receptors but no significant change in the mean dissociation constant, although two patients also had a clearly increased affinity.

2. Responses of Striatal Dopamine Receptors to Levodopa Treatment

Among the parkinsonian patients treated with levodopa there was a group of five patients who showed low specific binding of ^{3}H-spiroperidol and another group of five patients with high binding. In the former patients the binding of ^{3}H-spiroperidol was significantly reduced in the caudate nucleus and putamen but not in the pallidum, as compared with controls or patients with high binding (Table 2). According to Scatchard analysis, in

TABLE 1

Specific ^{3}H-spiroperidol binding (fmols/mg protein) in parkinsonian and control striatum. Mean±SEM.

Group	Number of patients	Caudate nucleus	Putamen	Pallidum	Receptor density x)
Controls					
1. Without neuroleptics	29	129±10	133±10	33± 3	101±14
2. With neuroleptics	9	293±13	289±22	81±10	306±36
Parkinsonian patients					
Without levodopa					
Sub-group I	5	53± 7	52± 7	41±12	55± 4
Sub-group II	10	173±18	228±25	66± 7	205±39
With neuroleptics	7	294±60	339±77	124±38	199±48
Control 1 – Sub-group I		p < 0.001	p < 0.001		p < 0.01
Control 1 – Sub-group II		p < 0.05	p < 0.01		p < 0.05
Control 1 – With neuroleptics		p < 0.05	p < 0.05	p < 0.001	
Sub-group I – Sub-group II		p < 0.001	p < 0.001	p < 0.05	p < 0.01

x) Dopamine-receptor density was assessed as maximum specific ^{3}H-spiroperidol binding derived from Scatchard analyses.

305

TABLE 2

Effect of levodopa treatment on specific ^{3}H-spiroperidol binding (fmols/mg protein) in parkinsonian and control striatum. Mean ± SEM.

Group	Number of patients	Caudate nucleus	Putamen	Pallidum	Receptor density x)
Controls					
1. Without neuroleptics	29	129±10	133±10	30±3	101±14
2. With neuroleptics	9	293±13	289±22	81±10	306±36
Parkinsonian patients					
With levodopa					
Sub-group I	5	67±12	69±12	46±12	53±14
Sub-group II	5	138±25	195±30	39±8	152±34
With neuroleptics	12	263±20	301±24	97±16	276±63
Control 1 – Control 2		p< 0.001	p< 0.001	p< 0.001	p< 0.001
Control 1 – Sub-group I		p< 0.001	p< 0.001		p< 0.05
Control 1 – With neuroleptics		p< 0.001	p< 0.001	p< 0.001	p< 0.01
Sub-group I – Sub-group II		p< 0.05	p< 0.01		p< 0.05

x) Dopamine-receptor density was assessed as maximum specific ^{3}H-spiroperidol binding derived from Scatchard analyses.

306

patients with low binding there was a significant decrease in
the number of receptors but no significant change in affinity.
On the other hand, in levodopa-treated patients with high
binding, neither the specific binding of ^{3}H-spiroperidol nor the
number of dopamine receptors differed significantly from those
of the controls (Table 2).

Analysis of clinical variables (Table 3) showed that in
patients with a high binding of ^{3}H-spiroperidol (Sub-group II)
the disability of the patients was less severe. Moreover, the
mean duration of the disease and the levodopa treatment was
somewhat, but not significantly, longer in the patients with low
binding. The mean daily dose of levodopa was similar in both
groups. On the other hand, as shown in Table 4, still before
death these patients with a high binding of ^{3}H-spiroperidol had
more therapeutic response to levodopa than those with a low
binding of ^{3}H-spiroperidol (Sub-group I), all of whom had a
deteriorating response to levodopa, only one still gaining
minimal beneficial effect. One patient in the first sub-group,
who was not bedridden, had suffered from daily freezing episodes. Similar on-off phenomena and dyskinesias appeared also in
two patients with a high binding of ^{3}H-spiroperidol. Two patients in both groups had psychotic episodes. On the other hand,
the interval between death and autopsy, 12 ± 6 hours, was
shorter than in patients with high binding, 32 ± 13 hours, but
the mean deep-frozen storage time of the brain samples was
similar in both groups.

In levodopa-treated patients who had received neuroleptic
medication for psychotic episodes there was significantly increased ^{3}H-spiroperidol binding in the caudate nucleus, putamen
and pallidum (Table 2). Scatchard analysis showed that the
number of receptors was significantly ($P < 0.01$) increased in
these patients (Table 2). There was no significant change in the
mean value of the dissociation constant, although in five patients there was also a clear-cut increase in affinity. Clinically, there were more involuntary movements in these patients
than in others treated with levodopa. Moreover, all but one of
these patients showed a deteriorating response to levodopa
treatment, but six patients were still deriving a beneficial
effect (Table 4). Other clinical variables (Table 3) did not
differ significantly from those of other levodopa-treated
patients.

3. Dopamine Receptors in the Nucleus Accumbens and Limbic Cortex

Table 5 shows that in patients with high ^{3}H-spiroperidol
binding in the striatum there was also increased binding in the
nucleus accumbens but not in the limbic cortex of parkinsonian
patients who had not had levodopa treatment. Furthermore, treatment with neuroleptics increased ^{3}H-spiroperidol binding in the

307

TABLE 3

Main clinical characteristics of parkinsonian patients treated with levodopa. Mean±SEM.

Group	Number of patients	Age (years)	Duration of disease (years)	Disability[x]			Levodopa treatment	
				III	IV	V	Duration (years)	Daily dose (mg)
1. Sub-group I (low binding)	5	70 ± 2	9 ± 3	0	1	4	3.6 ± 0.8	1260 ± 481
2. Sub-group II (high binding)	5	71 ± 3	6 ± 2	3	2	0	2.1 ± 0.9	865 ± 416
3. With neuroleptics	12	72 ± 2	11 ± 2	2	7	3	4.9 ± 0.7	929 ± 282

x) According to Hoehn and Yahr (1967).

TABLE 4

Main clinical responses of parkinsonian patients to levodopa at the time of death.

Group	Number of patients	Improvement			On-off effects	Dyskinesias	Psychotic episodes
		Mo.	Mi.	No			
Sub-group I (low binding)	5	0	1	4	1	0	2
Sub-group II (high binding)	5	2	1	2	2	2	2
With neuroleptics	12	1	5	6	5	9	12

Values are number of patients. Mo. = moderate, Mi. = minimal nucleus accumbens. On the other hand, in levodopa-treated patients, the specific binding of ^{3}H-spiroperidol did not differ significantly from that of controls, whereas treatment with neuroleptics increased ^{3}H-spiroperidol binding significantly in these levodopa-treated patients (Table 5).

4. Relationship between Parkinsonian Clinical Variables and Changes in Striatal Dopamine Receptors

The relationships between the clinical variables of the parkinsonian patients and changes in the striatal dopamine receptors are summarized in Table 6. The severity of the disease and a deteriorating response to levodopa seem to be mostly related to a decrease in the number of striatal dopamine receptors. On the other hand, dyskinesias, the daily fluctuations in performance and psychotic episodes together with neuroleptic medication were especially associated with an increase in the number of striatal dopamine receptors.

5. Dopamine Receptor Basis of Therapeutic Responses to Dopamine Receptor Agonists

Levodopa as a replacement therapy does not modify the natural course of Parkinson's disease. It merely improves parkinsonian symptoms but all the time while the treatment is going on, a progressive loss of substantia nigra neurons takes place leading to a considerable dopamine deficiency in the striatum (Rinne, 1978). However, there is a large innervation overlap with respect to the dopaminergic innervation in the striatum. Therefore a partial degeneration of the substantia nigra dopamine neurons leads to a greater loss of presynaptic dopamine neurons than postsynaptic dopamine receptors sites in the striatum. Comparative studies of striatal dopamine concentrations and dopamine receptors in the post-mortem parkinsonian brains have shown that this is really the case (Lee et al., 1978; Rinne et al., 1979). Thus in advanced parkinsonian patients who had lost the beneficial response to levodopa, there are still enough dopamine receptors in the striatum for drugs stimulating the dopamine receptors directly to improve the parkinsonian disability.

Indeed, recent evidence indicates that such agonists could be beneficial in the treatment of patients with Parkinson's disease (Calne et al., 1974, 1978; Cotzias et al., 1976; Lieberman et al., 1976, 1979; Parkes et al., 1976; Grøn, 1977; Rinne et al., 1977; Jansen, 1978; Rinne and Marttila, 1978; Fahn et al., 1979). In our long-term follow-up study it was shown that the additional bromocriptine therapy had a significant beneficial effect on the parkinsonian disability (Table 7) and the on-off phenomena (Table 8) of patients with a deteriorating levodopa response.

TABLE 5

Specific ^{3}H-spiroperidol binding (fmols/mg protein) in the nucleus accumbens and limbic cortex of parkinsonian patients and controls. Mean $\pm$ SEM.

Group	Nucleus accumbens	Limbic cortex
Controls		
1. Without neuroleptics	72 + 22 (4)x	108 + 15 (8)
2. With neuroleptics	174 + 44 (4)	174 + 15[2] (5)
Parkinsonian patients		
1. Without levodopa (Sub-group II)	136 + 8[1] (3) 193 (1)	129 + 5 (4)
2. Without levodopa and neuroleptics		171 + 82 (3)
3. With levodopa (Sub-group II)	113 + 18 (2)	120 + 20 (3)
4. With levodopa and neuroleptics	242 + 13[3] (4)	130 + 10 (5)

x) Number of patients
1) p< 0.05   as compared to controls without neuroleptics
2) p< 0.01
3) p< 0.001

TABLE 6

Relationship between clinical variables of the parkinsonian patients and changes in striatal dopamine receptors.

Variable	Striatal dopamine receptors	
	Decrease in the number	Supersensitivity
1. Disability of the patients	++	+
2. Loss of levodopa response	++	+
3. Dyskinesias	+	++
4. On-off phenomena	+	++
5. Psychotic episodes together with neuroleptics	0	++

TABLE 7

Bromocriptine-induced improvement (%) of parkinsonian patients with on-off phenomena during chronic levodopa treatment. Dosage (mg) of levodopa and bromocriptine. Mean ± SEM.

Variable	Duration of treatment (months)			
	1 (n=44)	3 (n=41)	6 (n=38)	12 (n=27)
Improvement				
Total disability	14 ± 2	20 ± 2	15 ± 3	14 ± 3
Tremor	36 ± 5	48 ± 5	39 ± 6	51 ± 6
Rigidity	12 ± 2	15 ± 3	10 ± 3	11 ± 4
Hypokinesia	12 ± 2	14 ± 3	8 ± 4	3 ± 6
Dosage				
Levodopa	920 ± 106	771 ± 52	703 ± 55	687 ± 67
Bromocriptine	23.8 ± 2.1	33.0 ± 4.3	30.0 ± 2.8	34.4 ± 4.8
	(2.5 - 70)x)	(10 - 160)	(5 - 80)	(5 - 120)

x) Range

TABLE 8

Responses (% of patients) of on-off disturbances to bromocriptine during long-term treatment with levodopa and bromocriptine.

Degree of improvement	Duration of treatment (months)			
	1 (n=41)x)	3 (n=36)	6 (n=33)	12 (n=27)
I. No improvement	13	11	18	26
II. Improvement	87	89	82	74
1. Minimal	20	33	37	37
2. Moderate	49	28	33	30
3. Marked	18	28	12	7

x) Number of patients suffering from on-off disturbances.

314

CONCLUSIONS

1. In Parkinson's disease there is not only a degeneration of dopaminergic substantia nigra neurons but also significant changes in the striatal dopamine receptors.

2. There seems to be at least two sub-groups of parkinsonian patients with regard to the behaviour of striatal dopamine receptors:

        I. A loss of dopamine receptors

        II. An increase in the number of dopamine receptors

3. In levodopa treated patients there was either a decreased or normal number of dopamine receptors but not a supersensitivity.

4. Behaviour of dopamine receptors in the nucleus accumbens was similar to those in the striatum.

5. Behaviour of dopamine receptors in the limbic cortex seems to differ from those in the striatum.

6. Clinically, the patients with a low number of striatal dopamine receptors were more disabled and had lost the beneficial response to levodopa.

7. In advanced parkinsonian patients with a deteriorating response to levodopa, there are still enough dopamine receptors in the striatum for drugs stimulating the dopamine receptors directly to improve the parkinsonian disability.

8. Bromocriptine seems to be a significant and valuable adjuvant therapy to levodopa in parkinsonian patients with a deteriorating response and/or the on-off phenomena.

SUMMARY

Striatal dopamine receptors were studied in 44 patients with Parkinson's disease by the radioligand-binding technique using ^{3}H-spiroperidol. The specific binding of ^{3}H-spiroperidol was either significantly increased or reduced in the caudate nucleus and putamen of parkinsonian patients without levodopa therapy. Scatchard analysis showed that there were corresponding changes in the receptor number, but no significant changes in the mean dissociation constant. The increased binding of ^{3}H-spiroperidol in the basal ganglia was also found in parkinsonian patients suffering from psychotic episodes and treated with neuroleptic drugs. Normal and low binding of ^{3}H-spiroperidol was found in patients treated with levodopa. Behaviour of dopamine receptors in the nucleus accumbens was similar to those in the striatum. However, the dopamine receptors in the limbic cortex seem to differ from those in the striatum.

Clinically, the patients with low binding of ^{3}H-spiroperidol in the striatum were more disabled and had lost the beneficial response to levodopa. Thus in Parkinson's disease in some patients a denervation supersensitivity seemed to develop and in some others a loss of postsynaptic dopamine receptor sites in the neostriatum. The latter alteration may contribute to the decreased response of parkinsonian patients to chronic levodopa therapy. However, in patients with a deteriorating response to levodopa, there seem to be still enough dopamine receptors in the striatum for drugs stimulating the dopamine receptors directly to improve the parkinsonian disability. Bromocriptine seems to be a significant and valuable adjuvant therapy to levodopa in parkinsonian patients with a deteriorating response and/ or the on-off phenomena.

ACKNOWLEDGEMENTS

The study was supported by a grant from the Sigrid Jusélius Foundation.

REFERENCES

Bernheimer, H., Birkmayer, W., Hornykiewicz, O., Jellinger, K. and Seitelberger, F. (1973). Brain dopamine and the syndromes of Parkinson and Huntington. J. Neurol. Sci., 20, 415-455.

Calne, D.B., Teychenne, P.F., Claveria, L.E., Eastman, R., Greenacre, J.K. and Petrie, A. (1974). Bromocriptine in parkinsonism. Br. Med. J., 4, 442-444.

Calne, D.B., Plotkin, C., Williams, A.C., Nutt, J.G., Neophytides, A. and Teychenne, P.F. (1978). Long-term treatment of parkinsonism with bromocriptine. Lancet I, 735-738.

Cotzias, G.C., Papavasiliou, P.S., Tolosa, E.S., Mendez, J.S. and Bell-Midura, M. (1976). Treatment of Parkinson's disease with apomorphines. N. Engl. J. Med., 294, 567-572.

Ehringer, H. and Hornykiewicz, O. (1960). Verteilung von Noradrenalin und Dopamine (3-hydroxytyramin) im Gehirn des Menschen und ihr Verhalten bei Erkrankungen des extrapyramidalen Systems. Klin. Wschr. 38, 1236-1239.

Fahn, S., Cote, L.J., Snider, S.R., Barrett, R.E. and Isgreen, W.P. (1979). The role of bromocriptine in the treatment of parkinsonism. Neurology 29, 1077-1083.

Grøn, U. (1977). Bromocriptine versus placebo in levodopa treated patients with Parkinson's disease. Acta neurol. Scand., 56, 269-273.

Jansen, E.N.H. (1978). Bromocryptine in levodopa response-losing parkinsonism. Eur. Neurol. 17, 92-99.

Lee, T., Seeman, P., Rajput, A., Farley, I.J. and Hornykiewicz, O. (1978). Receptor basis for dopaminergic supersensitivity in Parkinson's disease. Nature 273, 59-61.

Lieberman, A., Zolfaghari, M., Boal, D., Hassouri, H., Vogel, B., Battista, A., Fuxe, K. and Goldstein, M. (1976). The antiparkinsonian efficacy of bromocriptine. Neurology 26, 405-409.

Lieberman, A., Estey, E., Kupersmith, M., Gopinathan, G. and Goldstein, M. (1977). Treatment of Parkinson's disease with lergotrile mesylate. J. Amer. Med. Ass., 238, 2380-2382.

Lieberman, A., Dziatolowski, M., Kupersmith, M., Serby, M., Goodgold, A., Korein, J. and Goldstein, M. (1979). Dementia in Parkinson's disease. Ann. Neurol. 6, 355-359.

Lloyd, K.G., Davidson, L. and Hornykiewicz, O. (1973). Metabolism of levodopa in the human brain. In Advances in Neurology, (ed. D.B. Calne), Raven Press, New York.

Parkes, J.D., Marsden, C.D. and Donaldson, I. (1976). Bromocriptine treatment in Parkinson's disease. J. Neurol. Neurosurg. Psychiat., 39, 184-193.

Reisine, T.D., Fields, J.Z. and Yamamura, H.I. (1977). Neurotransmitter receptor alterations in Parkinson's disease. Life Sci., 21, 335-344.

Riederer, P. and Wuketich, St. (1976). Time course of nigrostriatal degeneration in Parkinson's disease. J. Neural Transm., 38, 277-301.

Rinne, U.K. (1978). Recent advances in research on Parkinsonism. Acta neurol. scand., 57, 77-113.

Rinne, U.K. and Sonninen, V. (1973). Brain catecholamines and their metabolites in Parkinsonian patients. Treatment with levodopa alone or combined with a decarboxylase inhibitor. Arch. Neurol., 28, 107-110.

Rinne, U.K. and Marttila, R.J. (1978). Brain dopamine receptor stimulation and the relief of Parkinsonism. Relationship between bromocriptine and levodopa. Ann. Neurol., 4, 263-267.

Rinne, U.K., Sonninen, V. and Hyyppä, M. (1971). Effect of L-Dopa on brain monoamines and their metabolites in Parkinson's disease. Life Sci., 10, 549-557.

Rinne, U.K., Sonninen, V., Riekkinen, P. and Laaksonen, H. (1974). Post-mortem findings in parkinsonian patients treated with L-Dopa: Biochemical considerations. In Current Concepts in the Treatment of Parkinsonism, (ed. M.D. Yahr), pp. 211-233, Raven Press, New York.

Rinne, U.K., Marttila, R.J. and Sonninen, V. (1977). Brain dopamine turnover and the relief of Parkinsonism. Arch. Neurol., 34, 626-629.

Rinne, U.K., Sonninen, V. and Laaksonen, H. (1979). Responses of brain neurochemistry to levodopa treatment in Parkinson's disease. In Advances in Neurology, (eds. L.J. Poirier, T.L. Sourkes and P.J. Bédard), 24, 259-274, Raven Press, New York.

A SUMMARY

Earl Usdin
National Institute of Mental Health Rockville, MD

In reading the summary in volumes such as this, I often wonder why it is included - is it that the Editor assumes that I may not have been able to understand what I have read or, Heaven forbid, that I may have skipped over a gem? My objective here is not to summarize the contents of the book, but rather to summarize the contents of the meeting which served as the basis for this volume. Actually I shall not even do that, but rather I shall mention some of the highlights of the meeting.

Thus, as an example, Arvid Carlsson, in a fairly heated discussion, asked the question: is there a nucleus accumbens in human brain? Arvid, possibly acting as a Devil's Advocate, indicated that he had no problem with the nucleus accumbens in brains of laboratory animals, that in these the nucleus was a visible, identifiable entity, but that this was not true for man. Oleh Hornykiewicz insisted that there is a nucleus accumbens in the human brain, that morphological identification of a locus is not the only criterion that should be used, that one can and should use pharmacological, enzymatic and other criteria.

Another paper which produced a good bit of discussion - and controversy - was Phil Seeman's. Here the provocative question was: Considering the relatively enormous concentration of dopamine required to affect it, why should Seeman's so-called D_1 receptor be considered as a D (i.e., dopamine) receptor? Phil's reply was that his three D receptors (D_1, D_2, D_3) were considered as dopamine receptors since each was more sensitive to dopamine than to any other neurotransmitter, even though their sensitivities to dopamine varied by orders of magnitude from one receptor type to another.

The use of fairly recently introduced techniqes such as positron emission tomography (PET) was discussed by several speakers. Steve Garnet, e.g., uses ^{18}F-fluoroDopa in his *in vivo* investigation of

intracerebral dopamine metabolism. The combination of PET and receptor assays (who could ask for anything more?) showed that only
about 1/5 of the Dopa extracted by the brain from blood is available for dopamine synthesis. Serotonin studies were covered by
Wolfgang Wesemann and GABA studies by Ken Lloyd. Toshi Nagatsu
discussed the use of his newly developed biopterin radioimmunoassay to show that there is a deficiency of biopterin in the brains
of patients with Parkinson's Disease and related striatal degenerative diseases.

Although I do not want to get caught doing exactly what I pro-
tested in my first sentence, I should like to mention a few specific items, such as Merton Sandler's presentation on dopamine-re-
lated enzymes in normal and pathological brains. In the striatum
and nucleus accumbens (assuming that Oleh is right in insisting
that this structure exists), dopamine is oxidized predominantly by
MAO-B rather than MAO-A. Merton also discussed the metabolism of
dopamine by dopa transaminase and by phenolsulfotransferase (but
this is a topic for another meeting).

Several of the speakers emphasized postmortem stability. Thus,
Ted Bird found that neuropeptides and various other moieties studied
in Huntington's Disease brains were quite stable, particularly in
the basal ganglia. Dr. Emson agreed that there is great stability
of peptides in postmortem human brain, including substance P, met-
enkephalin, vasoactive intestinal polypeptide, cholecystokinin, etc.
David Bowen found that choline acetyltransferase was quite stable
in postmortem human brain tissue and is a good marker for Alzheimer's disease.

Tim Crow indicated that he believes his results with brain tissue obtained from "non-medicated" (the quotes reflect my bias, not
Tim's) schizophrenics demonstrate that schizophrenia is not caused
by any of the following: norepinephrine neuron degeneration; deficit of MAO, GAD, GABA or serotonin; overactivity of dopamine neurons.

One issue which is becoming less and less of an issue with the
passage of time (or the passage of conferences) is that in most
neurological and psychiatric diseases the pathology is not the result of problems with a single neurotransmitter system; rather it
is a question of balance among a number of systems. Thus, Dr. Birkmayer stressed this balance hypothesis not only in the etiology
of Parkinsonism but also in various behavioral disorders. The involvement of multiple neurotransmitters was also emphasized by Bengt
Winblad; he noted that there is an age-related decrease in dopamine
in the nigrostriatal system, in norepinephrine in the hippocampus
and of serotonin in the gyrus cinguli. However, in the medulla oblongata, the level of serotonin increases with age - but the medulla oblongata is an unusual area: MAO-B activity decreases with age
here, even though it increases in all other brain areas so far examined.

Oleh Hornykiewicz and many of the other speakers emphasized that the results obtained in studies of human brain postmortem tissue depended not only on the patient (age, sex, prior diseases, medication history) but also on postmortem status (sudden death vs prolonged coma, anoxia, hyperthermia) and also on such items as time between death and autopsy, postmortem dissection techniques, etc. (A number of the world's human brain bankers are attempting to reach a consensus on standardized techniques for dissection, etc.)

The symposium on transmitter biochemistry of human brain tissue confirmed the involvement of many transmitter systems rather than only one in neurological and psychiatric illnesses and also indicated the complex nature of receptors for specific neurotransmitters. More questions were raised than answered - which in our present state of ignorance seems to be the hallmark of a successful seminar.

INDEX

325

329

in CSF of coma patients 157
in hepatic encephalopathies 146, 147, 148, 158
in Parkinson's disease 276
TRP-2,3-dioxygenase
cerebral ischemia 35
tryptophan-2,3-dioxygenase
in hepatic encephalopathy 146, 147, 164
tryptophan hydroxylase
cerebral ischemia 26, 35
tyrosine
death-authpsy influence 53
in hepatic failure 171
in Parkinson's disease 273
tyrosine hydroxylase
cerebral infarction 25
cerebral ischemia 26, 35
encephalopathy vascular 34
in hepatic failure 171
oxygen dependence 174
in Parkinson's disease 281-285, 293, 294, 295
in striatal degeneration 294, 295
L-valine 143
action in metabolic encephalopathies 144
and ammonia detoxification 169, 170
and citric acid cycle
competition at the blood brain barrier 149
therapy in hepatic coma 169, 170
vasoactive intestinal peptide
in human brain 223, 227, 229
in Huntington's chorea 212, 214, 230

SCIENCE AS PSYCHOLOGY

Science as Psychology examines the complexity and richness of scientific research by illustrating how social relationships, emotion, culture, and identity are implicated in the problem-solving practices of laboratory scientists. In this study, the authors gather and analyze interview and observational data from innovation-focused laboratories in the engineering sciences to show how the complex practices of laboratory research scientists provide rich psychological insights and how a better understanding of science practice facilitates the understanding of human beings more generally. The study focuses not on dismantling the rational core of scientific practice, but on illustrating how social, personal, and cognitive processes are intricately woven together in scientific thinking. The authors argue that this approach illustrates a way of addressing the integration problem in science studies – how to characterize the fluid entanglements of cognitive, affective, material, cultural, and other dimensions of discovery and problem solving. Drawing on George Kelly's "person-as-scientist" metaphor, the authors extend the implications of this analysis to general psychology. This book is thus a contribution to science studies, the psychology of science, and general psychology.

Lisa M. Osbeck is Associate Professor of Psychology at the University of West Georgia. She was a visiting research scientist in the School of Interactive Computing at the Georgia Institute of Technology during the year that most of this book was written. Dr. Osbeck is the recipient of the 2005 Sigmund Koch Award for Early Career Contributions to Psychology bestowed by the Society for Theoretical and Philosophical Psychology. She has served on the executive committee of the Society for Theoretical and Philosophical Psychology and is on the editorial boards of *Theory & Psychology*, the *Journal of Theoretical and Philosophical Psychology*, and the *International Journal for the Psychology of Science and Technology*.

Nancy J. Nersessian is Regents' Professor of Cognitive Science at Georgia Institute of Technology. She is a pioneer in the area of cognitive studies of science and technology, a former Chair and a Fellow of the Cognitive Science Society, a member of the governing board of the Philosophy of Science Association, and a founding member of the International Society for the Psychology of Science. Dr. Nersessian has received grants and fellowships from the National Science Foundation and the National Endowment for the Humanities and has held several residential fellowship positions, most recently at the Radcliffe Institute for Advanced Study at Harvard University. She is the author of numerous publications on the creative research practices of scientists and engineers, including the recent book *Creating Scientific Concepts*.

Kareen R. Malone is Professor of Psychology at the University of West Georgia and is also on the Women's Studies Faculty. She is associate editor of *Theory & Psychology*; is on the editorial board of *Psychoanalysis, Society, and Culture*; is a Fellow of the American Psychological Association; and served as 2010 President of the Society for Theoretical and Philosophical Psychology. Dr. Malone has coedited three volumes of Lacanian psychoanalysis; and is the author of numerous chapters, articles, and reviews in the areas of Lacanian psychoanalysis; gender, race, and science; feminism and the epistemology of psychology; cognitive psychology; and subjectivity.

Wendy C. Newstetter is Director of Learning Sciences Research in the Wallace H. Coulter Department of Biomedical Engineering at the Georgia Institute of Technology. She works with faculty at Georgia Tech and throughout the United States with Project Kaleidoscope to create and develop more effective science, math, and engineering educational environments informed by learning and cognitive science research. Dr. Newstetter has been published in the *Journal of Engineering Education*, *Research in Engineering Design*, and the *Annals of Biomedical Engineering*. She is senior editor for the *Journal of Engineering Education*.